Closed Circuit Television

Closed Circuit Television

Third edition

Joe Cieszynski
IEng MIET Cert. Ed. LCGI

AMSTERDAM • BOSTON • HEIDELBERG • LONDON • NEW YORK • OXFORD
PARIS • SAN DIEGO • SAN FRANCISCO • SINGAPORE • SYDNEY • TOKYO

ELSEVIER

Newnes is an imprint of Elsevier

Newnes

Newnes is an imprint of Elsevier
Linacre House, Jordan Hill, Oxford, OX2 8DP
30 Corporate Drive, Burlington, MA 01803

First edition 2001
Reprinted 2002
Second edition 2004
Reprinted 2004, 2005
Third edition 2007

British Library Cataloguing in Publication Data
A catalogue record for this book is available from the British Library

Library of Congress Cataloging-in-Publication Data
A catalog record for this book is available from the Library of Congress

ISBN-13: 978-0-7506-8162-9
ISBN-10: 0-7506-8162-4

For information on all Newnes publications
visit our website www.books.elsevier.com

Typeset by Charon Tec Ltd (A Macmillan Company), Chennai, India
www.charontec.com
Printed and bound in Great Britain

07 08 09 10 11   10 9 8 7 6 5 4 3 2 1

Contents

Preface

In the preface to the first edition I wrote that closed circuit television (CCTV) was a growth industry, the growth being very much a result of the impact of new technology. As I write the preface to this third edition of *Closed Circuit Television*, growth in the industry has continued, not only as a result of technological advances that continue to bring clearer images, more intelligent systems and lower equipment costs, but also because of the heightened awareness of risk that is prevalent in Western society today. There is a demand for everything from small, inexpensive systems to highly sophisticated systems covering many square miles.

And yet, like any high-technology installation, these systems will only function correctly if they are properly specified, installed, commissioned and maintained. Consequently, in addition to having an in-depth knowledge of CCTV principles and technology, the modern CCTV engineer is expected to be conversant with electrical and electronics principles, the latest digital and microprocessor principles, electrical installation practice, health and safety regulations, and telecommunications and network technologies.

Clearly no single textbook could provide a detailed coverage of all of these subjects, and it is the aim of this book to concentrate on CCTV principles and technology in order to provide the underpinning knowledge required by CCTV practitioners. Like the first two editions before it, this text will prove invaluable for those who are studying towards the City & Guilds Knowledge of Security and Emergency Alarm Systems (course 1852) and/or those who are working towards the NVQ level II or level III in CCTV installation and maintenance. On the other hand, this book is really intended for anyone who is involved with video signal processing and transmission, which naturally includes those who are practising in the industry and who wish to further their technical knowledge and understanding, but also includes anyone who uses closed circuit television for other applications such as surveying, medical, theatre production, etc.

As well as bringing the content of the second edition up to date, this third edition includes much new material on subjects such as the most recent (at the time of writing) video compression techniques, flat panel display technologies and structured (CAT 5/6) cable principles. A complete new chapter has been included to help engineers grasp the principles of modern networks and therefore have a better understanding of how to specify, set up and troubleshoot network CCTV systems.

It is my continued hope and wish that trainees and engineers alike will find this textbook a useful aid towards their personal development.

Joe Cieszynski

Acknowledgements

A text such as this would not be possible without the help and support of people in the industry and, since scripting the first edition of *Closed Circuit Television*, I have been assisted by a number of people, many of whom are specialists in their field. The people listed below have offered both their technical expertise and their time, for which I am very grateful.

Andrew Holmes of Data Compliance Ltd, with whom I have worked on numerous occasions, is a constant source of information. I am also indebted to Simon Nash of Sony, and Martin Kane, who have on many occasions provided me with information, help and guidance.

A thank you also to David Grant of ACT Meters and Gar Ning of NG Systems. Although their services were not called upon during the writing of this third edition, the marks that they made on the two previous editions remain.

I must also acknowledge the manufacturers who went out of their way to provide photographs and information for use in this book. Such support only helps in my quest to increase the level of knowledge and understanding of engineers in the industry. These manufacturers are acknowledged alongside their individual contributions.

A thank you to David Close for his sterling efforts in producing some of the photographic work which is used to illustrate video compression, and to Tim Morris of the University of Manchester for his much appreciated input into the video compression content in this book.

I would also like to thank my colleagues at PAC International Ltd for their support. In particular, Graeme Ashcroft for his proofreading of a number of portions of text, and Graham Morris and Steve Pilling who both spent much time proofreading the network theory, providing much appreciated feedback and suggested content.

As always, I am greatly indebted to my friend Ian Fowler for his input, which spans all three editions of this book. Once again he made himself available to discuss aspects of theory and technology and gave a lot of time to proofreading.

Finally, thanks again to David, Hannah, John and Ruth, my four (grown up) children, for their patience and support during the writing of this edition, and to Linda, my terrific wife, for her continued support.

1 The CCTV industry

The term 'closed circuit' refers to the fact that the system is self-contained, the signals only being accessible by equipment within the system. This is in contrast to 'broadcast television', where the signals may be accessed by anyone with the correct receiving equipment.

The initial development of television took place during the 1930s, and a number of test transmissions took place in Europe and America. In the UK these were from the Alexandra Palace transmitter in London. The outbreak of World War II brought an abrupt end to much of the television development, although interestingly transmissions continued to be made from occupied Paris using an experimental system operating from the Eiffel Tower; The Nazi propaganda machine was very interested in this new form of media.

Ironically the war was to give television the boost it needed in terms of technology development because in the UK it seemed as if every scientist who knew anything about radio transmission and signalling was pressed into the accelerated development programme for radar and radio. Following the war many of these men found themselves in great demand from companies eager to renew the development of television.

Early black and white pictures were of poor resolution; however, the success of the medium meant that the money became available to develop new and better equipment, and to experiment with new ideas. At the same time the idea of using cameras and monitors as a means of monitoring an area began to take a hold but, owing to the high cost of equipment, these early CCTV systems were restricted to specialized activity, and to organizations that had the money to invest in such security. These systems were of limited use because an operator had to be watching the screen constantly. There was no means of recording video images in the 1950s, and motion detection connected to some form of alarm was the stuff of James Bond (only even he did not arrive until the 1960s!).

Throughout the 1960s and 1970s CCTV technology progressed slowly, following in the footsteps of the broadcast industry, which had the money to finance new developments. The main stumbling block lay in the camera technology, which depended completely on vacuum tubes as a pick-up device. Tubes were large, required high voltages to operate, were generally useless in low light conditions (although special types were developed – for a price), and were expensive. Furthermore, an early colour camera required three of these tubes. For this reason, for many years CCTV remained on the whole a low-resolution, monochrome system which was very expensive.

By the 1980s camera technology was improving, and the cost of a reasonable colour camera fell to a sum that was affordable to smaller businesses and organizations. Also, VHS had arrived. This had a serious impact on the CCTV industry because for the first time it was possible to record video images on equipment that cost well below £1000. For a number of years prior to this, CCTV could be recorded on monochrome reel-to-reel machines, but these were expensive and were not exactly user-friendly.

From the mid-1980s onwards television technology advanced in quantum leaps. New developments such as the CMOS microchip and charge-coupled device (CCD) chip brought about an increase in equipment capability and greatly improved picture quality, whilst at the same time equipment prices plummeted. Manufacturers such as Panasonic and Sony developed digital video recording machines, and although these were intended primarily for use in the broadcast industry (at £50 000 for a basic model the CCTV industry was not in a hurry to include one with every installation!), these paved the way for digital video signal processing in lower-resolution CCTV and domestic video products.

For many years, CCTV had to rely on its big brother – the broadcast industry – to develop new technologies, and then wait for these technologies to be downgraded so that they became affordable to customers who could not afford to pay £30 000 per camera and £1000 per monitor. However, the technology explosion that we are currently seeing is changing this. PC technology is rapidly changing our traditional ideas of viewing and recording video and sound, and much of this hardware is inexpensive. Also, whereas in the early years the CCTV industry relied largely on the traditional broadcast and domestic television equipment manufacturers to design the equipment, there are now a large number of established manufacturers that are dedicated to CCTV equipment development and production. These manufacturers are already taking both concepts and hardware from other electronics industries and integrating them to develop CCTV equipment that not only produces high quality images, but is versatile, allows easy system expansion, is user friendly, and can be controlled from anywhere on the planet without having to sacrifice one of its most valuable assets – which is that it is a closed circuit system.

The role of CCTV

So often CCTV is seen as a security tool. Well of course it is; however, it plays equally important roles in the areas of monitoring and control. For example, motorway camera systems are invaluable for monitoring the flow of traffic, enabling police, motoring organizations and local radio to be used to warn drivers of problems, and thus control situations. And yet in the case of a police chase, control room operators can assist the police in directing their resources. This same versatility applies to town centre CCTV systems.

CCTV has become an invaluable tool for organizations involved in anything to do with security, crowd control, traffic control, etc. Yet on the other hand the proliferation of cameras in every public place is ringing alarm bells among those who are mindful of George Orwell's book *Nineteen Eighty-Four*. Indeed, in the wrong hands, or in the hands of the sort of police state depicted in that book, CCTV could be used for all manner of subversive activity. In fact the latest technology has gone beyond the predictions of Mr Orwell. Face recognition systems, which generate an alarm as soon as it appears in a camera view, have been developed, as have systems that track a person automatically once they have been detected. Other equipment which can see through a disguise by using parameters that make up a human, such as scull dimensions and relative positions of extreme features (nose, ears, etc.), or the way that a person walks, is likewise under development. At the time of writing all such systems are still somewhat experimental and are by no means perfected, but with the current rate of technological advancement we can only be a few years away from this equipment being installed as standard in systems in town centres, department stores, night clubs and anywhere else where the authorities would like early recognition of 'undesirables'.

To help control the use of CCTV in the UK, the changes made to the Data Protection Act (DPA) in 1998 meant that images from CCTV systems were now included. Unlike the earlier 1984 Act, this had serious implications for the *owners* of CCTV systems as it made them legally responsible for the management, operation and control of the system and, perhaps more importantly, the recorded material or 'data' produced by their system. The Data Protection Act 1998 requires that all non-domestic CCTV systems are registered with the Information Commissioner. Clear signs must be erected in areas covered by CCTV warning people that they are being monitored and/or recorded. The signs must state the name of the 'data controller' for the system, and have contact details. When registering a system, the data controller must state its specific uses and the length of time that material will be retained. Recorded material must be stored in a secure fashion and must not be passed into the public domain unless it is deemed to be in the public interest or in the interests of criminal investigations (i.e., the display of images on police-orientated programmes).

In 2004 the Information Commissioner's Office published a revised document in the light of a court case where the definition of the 'information relating to an individual' was challenged. Although the case did not directly involve CCTV 'information', nevertheless there were implications for smaller CCTV systems in the UK. The document advised that some smaller CCTV systems are not covered by the DPA because the information contained in their recordings cannot be considered to relate to an individual. By definition, if the cameras are fixed (i.e., no PTZ capability), are not used to monitor staff members to observe their behaviour, and recorded information is only passed to a law enforcement body such as the police, then the system does not have to be registered under the DPA.

On 2 October 1998 the Human Rights Act became effective in the United Kingdom. The emphasis on the right to privacy (among other things) has strong implications for CCTV used by 'public authorities' as defined by the Act and system designers and installers should take note of these implications. Cameras that are capable of targeting private dwellings or grounds (even if that is not their real intention) may be found to be in contravention of the rights of the people living there. As such, those people may take legal action to have the cameras disabled or removed – an expensive undertaking for the owner or, perhaps, the installing company who specified the camera system and/or locations.

In relation to CCTV, the intention of both the Data Protection and Human Rights Acts is to ensure that CCTV is itself properly managed, monitored and policed, thus protecting against it becoming a law unto itself in the future.

The arguments surrounding the uses and abuses of CCTV will no doubt continue; however, it is a well-proven fact that CCTV has made a huge, positive impact on the lives of people who live under its watchful eye. It has been proven time and again that both people and their possessions are more secure where CCTV is in operation, that people are much safer in crowded public places because the crowd can be better monitored and controlled, and that possessions and premises are more secure because they can be watched 24 hours per day.

The CCTV industry

Despite what we have said about CCTV being used for operations other than security, it can never fully escape its potential for security applications because, whatever its intended use, if the police or any other public security organization suspect that vital evidence may have been captured on a video recording system, they will inspect the recorded material. This applies all the way down to a member of the public who, whilst innocently using a camcorder or a video recorder on a mobile phone, happens to capture either an incident or something relating to an incident. For this reason it is perhaps not surprising to hear that the CCTV industry is largely regulated and monitored by the same people and organizations that monitor the security industry as a whole.

The British Security Industry Association (BSIA) Ltd is the only UK trade association for the security industry that requires its members to undergo independent inspection to ensure they meet relevant standards. The BSIA's primary role is to promote and encourage high standards of products and services throughout the industry for the benefit of customers. This includes working with its members to produce codes of practice, which regularly go on to become full British/European standards. The BSIA also lobbies government on legislation that may impact on the industry and actively liaises with other relevant organizations, for example the Office of the Information Commissioner (in relation to the Data Protection Act) and the Home Office Scientific Development Branch (HOSDB). The BSIA also provides an invaluable service in producing technical literature and training materials for its members and their customers.

Inspectorate bodies are charged with the role of policing the installation companies, making sure that they are conforming to the Codes of Practice. Of course, a company has to agree to place itself under the canopy of an Inspectorate, but in doing so it is able to advertise this fact, and gives it immediate recognition with insurance companies and police authorities.

To become an approved installer a company must submit to a rigorous inspection by its elected Inspectorate. This inspection includes not only the quality of the physical installation, but every part of the organization. Typically the inspector will wish to see how documentation relating to every stage of an installation is processed and stored, how maintenance and service records are kept, how material and equipment is ordered, etc. In addition the inspector will wish to see evidence that the organization has sufficient personnel, vehicles and equipment to meet maintenance requirements and breakdown response times.

In some cases the organization is expected to obtain BS EN ISO 9002 quality assurance (QA) accreditation within two years of becoming an approved installer. At the time of writing there is no specific requirement that engineers working for an approved installation company hold a National Vocational Qualification (NVQ) in security and emergency systems engineering; however, this may well become the case in the future.

Another significant body is Skills for Security, the Standards Setting Body for the security business sector. Skills for Security incorporates many of the functions formerly undertaken by SITO (Security Industry Training Organization) as well as adopting a wider remit similar to Sector Skills Councils. In the UK, SITO were responsible for the development of training standards for the security industry and did much to raise those standards throughout the 1990s. They developed the NVQ levels II and III for electronic security systems, plus many other awards covering all sectors of the security industry. Skills for Security came into being in January 2005 and work closely with the industry to identify the training needs (both present and future) and develop programmes and qualifications that will meet these needs.

Awarding bodies such as City & Guilds and Edexcel play an important role in the security industry because it is they who devise the course syllabus and assessment criteria for the training and education of personnel working in the industry. The UK qualification for CCTV engineers is the City & Guilds NVQ level II or level III in Security and Emergency Alarm Systems. The City & Guilds also offer the underpinning knowledge test papers (course 1852) for the four disciplines relating to security and emergency systems engineering, these being CCTV, intruder alarm, access control and fire alarm systems. These awards are intended to contribute towards the underpinning knowledge testing for the NVQ level III award, although a candidate may elect to sit these tests without pursuing an NVQ. It must be stressed, however, that the 1852 award is not an alternative qualification to an NVQ, and a person holding only the 1852 certificates would not be deemed to be qualified until they had proven their competence in security systems engineering.

The awarding bodies appoint external verifiers whose role it is to check that NVQ assessment centres, be these colleges, training organizations or installation companies, are carrying out the assessments to the recognized standards.

The Home Office Scientific Development Branch (formerly the Police Scientific Development Branch – PSDB) plays a most significant role in CCTV. For many years the CCTV industry had no set means of measuring the performance of its systems in terms of picture quality, resolution and the size of images as they appear on a monitor screen. This meant that in the absence of any benchmarks to work to, each surveyor or installer would simply do what they considered best. This situation was not only unsatisfactory for the industry; potential customers were in a position where they had no way of knowing what they could expect from a system and, once it was installed, had no real redress if they were unhappy, because there was nothing for them to measure the system performance against.

The PSDB set about devising practical methods of defining and measuring such things as picture resolution and image size and, for example, in 1989 introduced the Rotakin method of testing the resolution and size of displayed images (see Chapter 13). They also developed methods of analysing and documenting the needs of customers prior to designing a CCTV system. This is known as an Operational Requirement (OR). HOSDB continue this work, providing much practical guidance on issues relating to the latest CCTV technologies such as watermarking of recorded video images, methods of archive retrieval, measurement of resolution of digital images, etc.

CCTV is a growth industry. It has proven its effectiveness beyond all doubt, and the availability of high-quality, versatile equipment at a relatively low cost has resulted in a huge demand for systems of all sizes. Within the industry there is a genuine need for engineers who truly understand the technology they are dealing with, and who have the level of underpinning knowledge in both CCTV and electronics principles that will enable them to learn and understand new technologies as they appear.

2 Signal transmission

A CCTV video signal contains a wide range of a.c. components with frequencies varying from 0 Hz up to anything in the order of 10 MHz. Furthermore, in addition to the a.c. components there is also an essential d.c. component which must be preserved throughout the signal transmission process if accurate brightness levels are to be maintained. Problems occur when engineers consider video signal transmission in the same terms as transmitting low-voltage d.c. or low-frequency mains voltage. When you consider that domestic medium wave radio is transmitted around 1 MHz, then it becomes clear that the 0–10 MHz video signal is actually going to behave in a similar manner to radio signals.

In this chapter we shall examine the peculiar behaviour of high-frequency signals when they are passed along various types of cables, and therefore explain the need for special cables when transmitting video signals, and the reasons for the limitations in each transmission medium.

CCTV signals

An electronically produced square wave signal is actually built up from a sinusoidal wave (known as the fundamental) and an infinite number of odd harmonics (odd multiples of the fundamental frequency). This basic idea is illustrated in Figure 2.1 where it can be seen that the addition of just

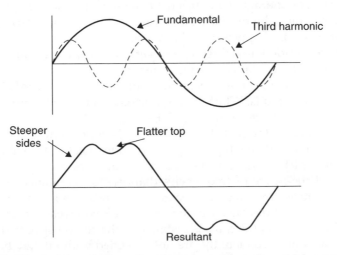

Figure 2.1 *Effect of the addition of odd harmonics to a sinusoidal waveshape*

the third harmonic component changes the appearance of the fundamental sine wave, moving it towards a square shape. Adding the fifth harmonic would have the effect of steepening the sides and flattening the top. In other words the waveshape becomes more square. Taking this to the extreme, adding an infinite number of odd harmonics would produce a waveshape that has perfectly vertical sides and a perfectly flat top.

If we reverse this process, i.e., begin with a square wave and remove some of the harmonic components using filters, then the corners of the square wave become rounded, and the rise time becomes longer. In other words, the square wave begins to return to its sinusoidal fundamental. This effect is illustrated in Figure 2.2.

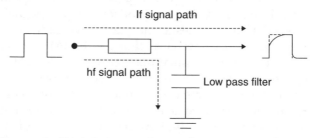

Figure 2.2 *Removal of high-frequency harmonic components increases the rise time and rounds the corners*

In Chapter 5 we shall be looking at the make-up of the video signal (Figure 5.13), and we will see that it contains square wave components. *It is the sharp rise times and right-angled corners in the video signal waveform which produce the high-definition edges and high-resolution areas of the picture.* If for any reason the signal is subjected to a filtering action resulting in the loss of harmonics, the reproduced picture will be of poor resolution and may have a smeared appearance. Now one may wonder how a video signal could be 'accidentally' filtered, and yet it actually occurs all of the time because all cables contain elements of resistance, capacitance and inductance, the three most commonly used components in the construction of electronic filter circuits. When a signal is passed along a length of cable it is exposed to the effects of these R, C, L components.

The actual effect the cable has on a signal is dependent on a number of factors which include the type and construction of cable, the cable length, the way in which bends have been formed, the type and quality of connectors and the range of frequencies (bandwidth) contained within the signal. This means that, with respect to CCTV installations, it is important that correct cable types are used, that the correct connectors are used for a given cable type, that the cable is installed in the correct specified manner and that maximum run lengths are not exceeded without suitable means of compensation for signal loss.

Different cable types are used for the transmission of CCTV video signals, and indeed methods other than copper cable transmission are employed. Both the surveyor and the installing engineer need to be aware of the performance and limitations of the various transmission media, as well as the installation methods that must be employed for each medium.

Co-axial cable

As stated earlier, the behaviour of high-frequency signals in a copper conductor is not same as that of d.c. or low frequencies such as 50/60 Hz mains, or audio, and specially constructed cables are required to ensure constant impedance across a range of frequencies. Furthermore, radio-frequency signals have a tendency to see every copper conductor as a potential receiving aerial, meaning that a conductor carrying an RF signal is prone to picking up stray RF from any number of sources; for example emissions from such things as electric motors, fluorescent lights, etc., or even legitimate radio transmissions. Co-axial cable is designed to meet the unique propagation requirements of radio-frequency signals, offering a reasonably constant impedance over a range of frequencies and some protection against unwanted noise pick-up.

There are many types of co-axial cable, all manifesting different figures for signal loss, impedance, screening capability and cost. The construction of a co-axial cable determines the characteristics for a particular cable type, the basic physical construction being illustrated in Figure 2.3.

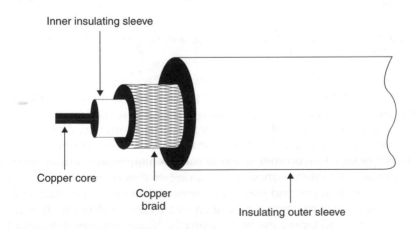

Figure 2.3 *Co-axial cable construction*

The signal-carrying conductor is the copper central core, which may be a solid copper conductor or stranded wire. The signal return path could be considered to be along the braided screen; however, as this is connected to the earth of a system, the signal may in practice return to its source via any number of paths. However, the screen plays a far more important role than simply to serve as a signal return path. It provides protection against *radio-frequency interference* (RFI). The way that it achieves this is illustrated in Figure 2.4, where it can be seen that external RF sources in close proximity of the cable are attracted to the copper braided screen, from where they pass to earth via the equipment at either end of the cable. Provided that the integrity of the screen is maintained at every point along the cable run from the camera to the monitor, there is no way that unwanted RF signals can enter either the inner core of the co-axial cable or the signal processing circuits in the equipment, which will themselves be screened, usually by the metal equipment casing.

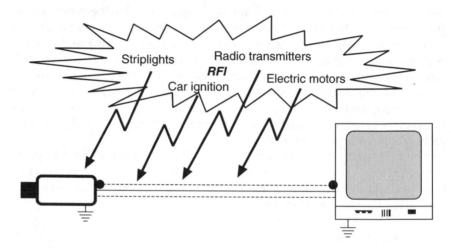

Figure 2.4 *RFI is contained by the copper screen, preventing it from entering the signal processing circuits*

Integrity of the screen is maintained by ensuring that there are no breaks in the screen at any point along the cable length, and that all connectors are of the correct type for the cable and have been fitted correctly. We shall consider connectors later in this chapter, but the issue of breaks in the screen is one which we need to consider. Co-axial cable is more than a simple piece of wire, and only functions correctly when certain criteria have been met in relation to terminations and joints. Under no circumstances should a joint be made by simply twisting a pair of cores together and taping them up before twisting and taping the two screens. Although this might appear to be electrically sound, it breaks all the rules of RF theory and, among other things, can alter the dynamic impedance and expose the inner core to RFI. All joins should be made using correctly a fitted connector (usually BNC) on each cable end, with a coupling piece inserted in between.

Where RFI is present in a video signal, it usually manifests itself as a faint, moving patterning effect superimposed onto the picture. The size and speed of movement of the pattern depends on the frequency of the interfering signal.

The inner sleeve of the co-axial cable performs a much more important function than simply insulation between the two conductors: it forms a dielectric between the conductors which introduces a capacitive element into the cable. This cable capacitance works in conjunction with the natural d.c. resistance and cable inductance to produce a *characteristic impedance* (Z_o) for the cable. One of the factors which governs the value of a capacitor is the type of dielectric (insulator) used between the plates, and co-axial cables of differing impedances are produced by using different materials for the inner core. This is why not all co-axial cables are suitable for CCTV applications, and why a connector designed for one cable type will not fit onto certain other types; the cable diameter varies depending upon the dielectric. The equivalent circuit of a co-axial cable is shown in Figure 2.5.

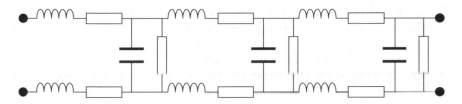

Figure 2.5 *Equivalent circuit of a co-axial cable, also known as a transmission line*

The characteristic impedance for a cable of infinite length can be found from the equation $Z_o = \sqrt{L/C}$. However, this concept is somewhat theoretical as we do not have cables of infinite length. On the other hand, for a co-axial cable to function as a transmission line with minimum signal loss and reflection (we will look at this is a moment), the termination impedance at both ends must equal the calculated characteristic impedance for an infinite length. Thus, if the characteristic impedance, Z_o, for a cable is quoted as being 75 Ω, then the equipment at both ends of the cable must have a termination impedance of 75 Ω.

If this is not the case a number of problems can occur. First of all signal loss may be apparent because of power losses in the transfer both to and from the cable. It can be shown that for maximum power transfer to occur between two electrical circuits, the output impedance of the first circuit must be equal to the input impedance of the second (Figure 2.6). If this is not the case, some power loss will occur. In our case the co-axial cable can be considered to be an electrical unit, and this is why all equipment connected to the cable must have a matching impedance.

Another problem associated with incorrect termination is one of *reflected waves*. Where a cable is not terminated at its characteristic impedance, not all of the energy sent down the line is absorbed by the load, and

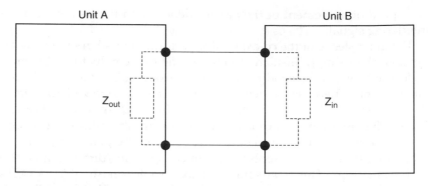

Figure 2.6 *Maximum power transfer only occurs when Z_{out} in Unit A is equal to Z_{in} in Unit B. (Assume that the connecting cables have zero impedance)*

because the unabsorbed energy must go somewhere, it travels back along the line towards its source. We now have a situation where there are two signals in the cable, the forward wave and the reflected wave. In CCTV, *reflected waves can cause ghosting, picture roll, and loss of telemetry signals.* However, these symptoms may not be consistent and may alter sporadically, leaving the unsuspecting service engineer chasing from one end of the installation to the other looking for what appears to be a number of shifting faults – and perhaps for no other reason than because a careless installation engineer has made a Sellotape-style cable connection in a roof space!

CCTV equipment is designed to have 75 Ω input and output impedances. This means that 75 Ω co-axial cable must always be used. Here again the installing engineer must be aware that not all co-axial cable has 75 Ω impedance, and 50 Ω and 300 Ω versions are common. For example, cable type RG-59 is a common 75 Ω co-axial cable used in CCTV installations. Cable type RG-58 looks very similar, but it is designed for different applications and has a characteristic impedance of 50 Ω. A CCTV installation using this cable would never perform to its optimum capability, if indeed it were able to perform at all.

Termination switches are included in CCTV equipment to ensure that there is a 75 Ω impedance at both ends of any co-axial cable network. This topic will be discussed in more detail in Chapter 7.

Up to now we have not taken into consideration the length of the co-axial cable. Over short distances the effects of C and R on the signal are small and can be ignored. However, as the cable length is increased, these components have an effect on the signal which is similar to a voltage drop along a d.c. supply cable, the main difference being that the filtering action of the cable results in greater losses at the higher signal frequencies. Figure 2.7 illustrates a typical co-axial cable frequency response. Cable losses are usually quoted in terms of dB per 100 m, at a given frequency. Manufacturers may quote figures for a range of frequencies; however, those quoted for around 5 MHz are the most significant to the CCTV engineer because, as seen from Figure 2.7, it is at the top end of the video signal frequency spectrum where the most significant losses occur.

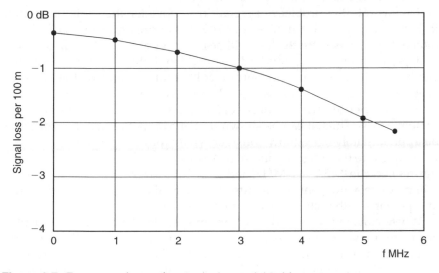

Figure 2.7 *Frequency losses in a typical co-axial cable*

Every cable employed in CCTV signal transmission has a specified maximum length, beyond which optimum system performance will only be maintained if additional equipment is installed. Typical specifications for the three most common co-axial cables employed in the CCTV industry are given in Table 2.1. The figures quoted for the maximum cable run length are those quoted in the BSIA Code Of Practice for Planning, Installation and Maintenance of CCTV Systems, document 109 issue 2 1991, and some variance with these figures may be noted when comparing different manufacturers' data; however, the installer will do well to heed the guidelines laid down in the BSIA document.

Table 2.1 *Co-axial cables commonly employed in the CCTV industry*

Cable type	Max run length	Impedance	Loss/100 m at 5 MHz
URM-70	250 m	75 Ω	3.31 dB
RG-59	350 m	75 Ω	2.25 dB
RG-11	700 m	75 Ω	1.4 dB

As a rule, monochrome signals tend to cope better with long cable runs than do colour signals. This is because a PAL colour signal contains a high-frequency 4.43 MHz colour subcarrier which is affected by the filtering action of the cable. However, even a monochrome signal contains h.f.

components and, bearing in mind the effects of h.f. filtering on a square wave (Figure 2.2), will therefore suffer some loss of resolution and signal level where the cable run length is excessive.

To illustrate the problem of signal loss, consider the cable illustrated in Figure 2.7. At 3 MHz the loss per 100 m is approximately -1 dB. Thus, over a distance of 350 m the loss will be in the order of -3.5 dB. In terms of voltage, assuming that a standard 1 Vpp video signal was injected into the cable, -3.5 dB represents an output voltage at the end of the cable of around 0.7 Vpp; a signal loss of 0.3 V. At 5 MHz the loss is in the order of -1.75 dB per 100 m; therefore over 350 m the loss in dBs will be approximately -6 dB. Thus, it can be shown that at 5 MHz the signal output will be approximately 0.5 V. Now consider what would happen if an installer were to ignore these figures and fit a 700 m length of this cable. The output figures become 0.45 V at 3 MHz, and 0.25 V at 5 MHz. At best such a signal will produce a low-contrast picture, more than likely with a loss of colour, and perhaps with picture roll due to the loss of sync pulses.

Where runs in excess of the maximum specified length for a particular cable are unavoidable, launch amplifiers and/or cable equalizers can be installed. The use of these can at least double the length of a cable run.

A launch amplifier is usually installed at the camera end of the cable where there is an available source of power, although there is a sound argument for installing it half way along a length of cable if a means of supplying power can be found. A typical launch amplifier response is shown in Figure 2.8 where it can be seen that the level of amplification is not uniform across the 0–5.5 MHz video signal bandwidth. The amplifier is designed to give extra lift to the higher frequencies where the greater losses occur.

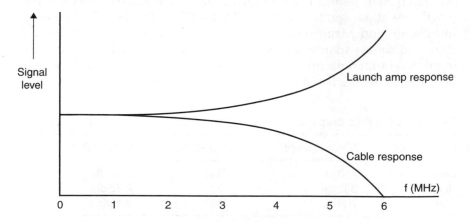

Figure 2.8 *A launch amplifier compensates for the filter action of the cable*

The amplifier usually has an adjustment to allow the gain to be set to suit the length of cable; the longer the cable the higher the gain setting. The idea is to set the output voltage level such that, after losses, a uniform

1 Vpp signal appears at the other end. In some cases the gain control is calibrated in cable lengths, and it is therefore necessary to have an approximate idea of the length of the run. Do not simply turn the control until a 'good, strong picture' appears on the monitor. This practice can lead to problems in relation to vertical hold stability where switchers or multiplexers are involved, and possibly a loss of picture resolution.

A cable equalizer is a form of amplifier, but it is designed to be installed at the output end of the cable (Figure 2.9). The problem with this is that the unit is having to process the signal once the losses have been incurred, and in boosting the signal levels it will also boost the background noise level, which will have risen in the absence of a strong signal. The advantage of using a cable equalizer is that it can be installed in the control room, which can be a real plus in cases where the camera is inaccessible. If the installer has a choice of which to use, a launch amplifier is usually preferable as it lifts the signal before losses occur, thus maintaining a better signal-to-noise ratio.

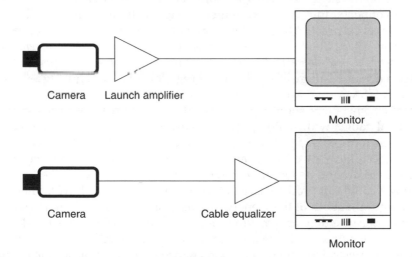

Figure 2.9 *Use of launch amplifiers and cable equalizers*

It is possible to employ more than one amplifier in cases where very lengthy cable runs are required. The idea is that these are placed at even distances along the cable such that, just as the signal would begin to deteriorate, another amplifier lifts it once again. This principle is shown in Figure 2.10 where it can be seen that the total cable loss is −33.75 dB, which is compensated for by the overall gain in the system of 36 dB.

All this sounds well and good, but it takes a highly experienced engineer with the correct equipment to be able to adjust the gain and response of all of these units to a point where a perfect, uniform 1 Vpp, 0–5.5 MHz video signal is obtained at the other end without any increase in noise level. And remember, once noise has been introduced into the signal, it will simply be boosted along with the signal in each subsequent amplifier.

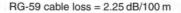

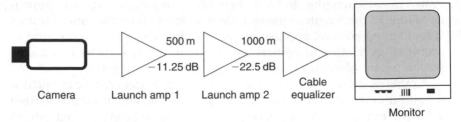

Figure 2.10 *Launch amp 1 gain = 12 dB. This compensates for the first 500 m of cable. Launch amp 2 gain = 12 dB. This compensates for 50% of the losses in the following l km of cable. Cable equalizer gain = 12 dB correcting for losses in the 1 km cable run*

Still on the subject of losses, it should be noted that every BNC (or other type) connector introduces an element of signal attenuation and reflection, and it is good practice to keep the number of joins in a cable to a minimum.

In the UK, all CCTV signal cable installation should comply with current codes of practice as laid down in BS 7671; Requirements for Electrical Installations, especially in relation to electrical segregation of low- and high-voltage cables. However, apart from the electrical safety issues surrounding segregation, installers should pay particular attention to the proximity of co-axial cables with mains power cables, in particular those carrying a high current, or supplying large numbers of fluorescent lights, heavy machinery, etc. Any current-carrying conductor produces an electromagnetic field around its length. Furthermore, high-frequency spikes passing along a cable can produce large electromagnetic pulses (EMP). Therefore it follows that both of these energy fields must surround all mains supply cables, because they are carrying high-frequency noise spikes in addition to the high current 50 Hz mains supply. Where co-axial cables are run parallel to mains cables, there is a good chance of the *electromagnetic interference* (EMI) penetrating the screen and superimposing a noise signal onto the video signal. Where this occurs, the displayed or recorded picture will suffer such effects as horizontal ripples rolling up or down, or random flashes when lights are switched or machinery operated.

Naturally the co-axial screen provides much protection against such noise ingression, but at best the screen will be no more than 95% effective; and many cables may have a much lower figure. To prevent noise ingression it is good practice to avoid long, close-proximity, parallel co-axial/mains supply cable runs wherever possible, maintaining at least 30 cm (12″) between cables. This may rule out using plastic segregated trunking because, although it offers electrical segregation, it does nothing to prevent the problems we have just outlined. Metal trunking provides screening against interference, and in cases where co-axial video cable must run through areas of high electrical noise, it is good practice to use steel trunking or conduit to minimize the chances of EMI compromising system performance.

Having looked at the construction of co-axial cable we know that the characteristic impedance depends, among other things, upon the capacitance of the cable, which is determined by the type and thickness of the inner insulating material. Therefore, should the inner sleeve become damaged by the cable being crushed, kinked, or filled with water, the characteristic impedance will alter, opening the system up to the inherent problems of signal loss and reflected waves. Putting this another way, installers should take care not to damage the cable during installation, and should not lay cables in places where they may easily be damaged at a later time. BNC connectors are not waterproof and were never intended for external use. Therefore where external connections are necessary, they should always be enclosed in a weatherproof housing. Once water enters a co-axial cable the capillary action may allow it to travel many metres along the cable, introducing all manner of undesirable picture effects, and very often these can be intermittent.

In order to prevent damage to the inner sleeve, co-axial cable should not have any severe bends. A rule of thumb is to ensure that *the radius of all bends is no tighter than five times the diameter of the cable*. For example, if the cable diameter is 6.5 mm, the radius of a bend should be at least 32.5 mm.

Ground loops

These occur when the earth (voltage) potential differs across the site. Because every item of mains powered equipment must be connected to earth, where the earth potentials differ, an a.c. 50 Hz current will flow through the low-impedance screen. The problem is illustrated in Figure 2.11 where a length of co-axial cable has a potential difference of +40 V between its ends. It naturally follows that a current will flow through the low-impedance co-axial screen which is bypassing the much higher impedance of the ground, which was the cause of the potential difference in the first place.

Differing ground potentials are very common, especially over long distances, and the problem can be further compounded when equipment at one end of a cable is connected to a different phase of the mains supply than that at the other end. The example in Figure 2.11 indicates a potential difference of 40 V; however, a difference of just 2–3 V is sufficient to cause problems.

When a ground loop current flows along a co-axial cable screen, because the centre core is referenced to the screen, a 50 Hz ripple is superimposed onto the video signal. This means that the brightness levels in the signal information are constantly moving at a rate of 50 Hz, and the effect on the monitor display is either a dark shadow or a ripple rolling vertically through the picture. This effect, known often as a *hum bar*, can also upset the synchronizing pulses, resulting in vertical picture roll.

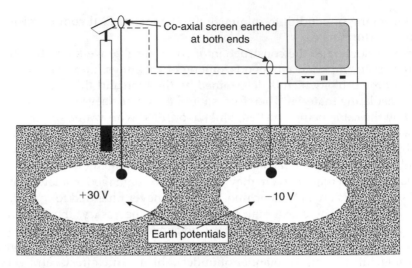

Figure 2.11a *A CCTV system where earth potentials differ*

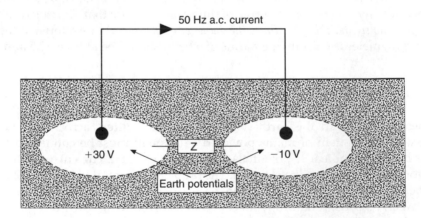

Figure 2.11b *Equivalent electrical circuit. The high-impedance earth path (Z) is by-passed by the low impedance of the co-axial screen*

It is possible to test for an earth potential problem during installation by taking an a.c. voltage measurement between the co-axial screen and the earth of the equipment to which it is to be connected. Under perfect earthing conditions, the reading should be 0 V. In practice it is usual to obtain a reading of at least a few hundred millivolts; however, in severe conditions potentials of 50 V or even greater are possible. In such cases it is not safe to assume that the problem is simply caused by differences in earth potential as there might actually be a serious fault in the earth circuit of the electrical supply, and if the CCTV installer himself is not a qualified electrician, he should report the potential fault to the appropriate persons, in writing, in order that a full inspection of the supply can be carried out.

There are various ways of avoiding or overcoming the problem of ground loops in a CCTV system. Avoidance is always the best policy, but is not always practical. Remember that ground loops occur because the system has more than one earth point, and these are at differing potentials. Therefore if 12 Vd.c. or 24 Va.c. cameras can be used, the only earth connection to the co-axial cable is at the control room end, and ground loops will not occur. This principle is illustrated in Figure 2.12. Other methods of avoidance are to employ twisted pair or fibre-optic cables, which we shall be looking at later in this chapter. However, fibre-optic cables are more expensive to install, and besides, the problem may not be identified until after co-axial cables have already been installed.

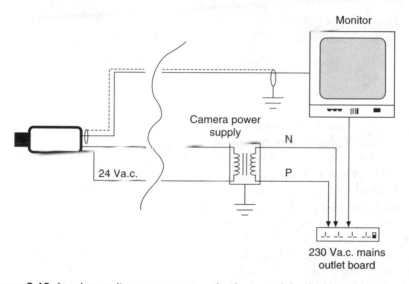

Figure 2.12 *In a low-voltage camera supply, the co-axial cable is only earthed at the monitor end*

Ground loop correction equipment is available. There are two types: transformer and optical. Transformer types are usually contained in a sealed metal enclosure which acts as screening. In order to provide ideal coupling of the broadband video signal, the internal circuits may contain more than just a transformer. Nevertheless, the principle behind these units is to break the co-axial cable earth circuit but still provide video signal transmission without affecting the integrity of the cable screen. The basic circuit operation is shown in Figure 2.13. In practice a single unit may contain two transformers, allowing two separate video circuits to be corrected. The unit can be installed at either end of the cable, although it is usually more convenient to locate it at the control room end.

It is worth noting that not all correction transformers perform to the same standard when it comes to broadband video signal coupling, and sometimes

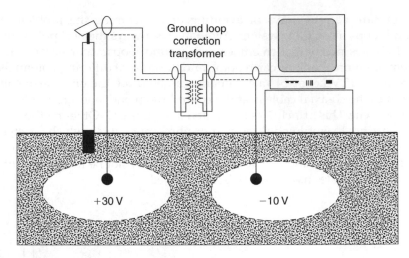

Figure 2.13 *Inclusion of a transformer breaks the 50 Hz current path through the co-axial screen*

a loss of resolution may be evident. Furthermore, where a transformer is not capable of coupling high frequencies, this can pose problems for certain types of telemetry control signal, resulting in a loss of telemetry to cameras which have a ground loop correction transformer included. As with any type of CCTV equipment, careful selection is important, and when you have found a product which performs satisfactorily, stay with it.

Optical correctors rely on opto-couplers to break the co-axial screen (Figure 2.14). The video signal is applied to a light-emitting diode which converts the varying voltage levels in the video signal into variations in light level. These in turn are picked up by a photodiode which converts the light signal back into a variable voltage. Units containing a number of individual inputs (typically 8 or 16) are available, and can be included with the

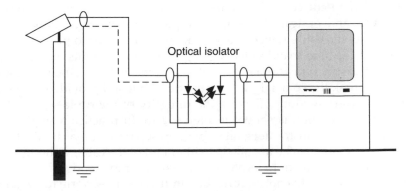

Figure 2.14 *Principle of a single channel opto-isolator*

control room equipment, acting as a buffer for each camera input. These are ideal for installations where it is anticipated at the planning stage that ground loops may pose a problem because it is known that cameras will either be connected across different phases of the mains supply, or will span a large geographical area. A multiple input ground loop corrector can be included in the initial quotation, thereby removing the problems of additional costs once the installation is underway.

Twisted pair cable

As the name implies, this cable comprises two cores which are twisted around each other. The number of twists per metre varies depending upon the quality of the cable, but a minimum of 10 turns per metre is recommended for CCTV video signal transmission applications – the more turns there are, the better the quality of the cable in terms of noise rejection.

This type of cable provides *balanced signal transmission* (as opposed to *unbalanced*, which is how co-axial cable functions). As illustrated in Figure 2.15, in a balanced transmission system, because the two conductors are twisted together, they are evenly exposed to any sources of electrical or magnetic interference present. Furthermore, the induced noise signals travel in the same direction along both conductors, whereas the video signal is travelling in opposite directions along each conductor (signal send and return).

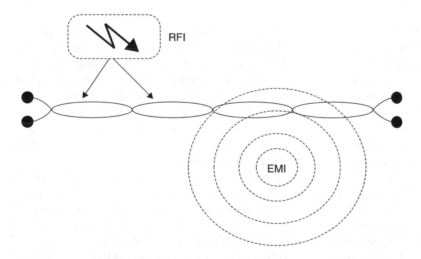

Figure 2.15 *Noise is induced equally into both conductors in a twisted pair*

A receiver unit containing an *operational amplifier (op-amp)* circuit is inserted at the output end of the cable for the purposes of noise cancellation. An op-amp has two inputs, and the conductors in the twisted pair are connected to these inputs. Because the noise signals are travelling in the same direction on both conductors, they are effectively applied to both op-amp inputs in the same phase. However, the action of the op-amp is such that the noise signals are added in antiphase and they are thus cancelled out. This noise-cancelling action is illustrated in Figure 2.16. The video signal, on the other hand, is only present on the 'send' conductor and is therefore only applied to the non-inverting input on the op-amp. Thus, the only signal present at the op-amp output will be the video signal.

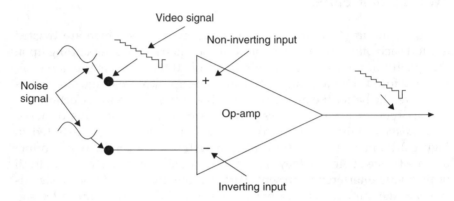

Figure 2.16 *Noise at the inverting input is added to that at the non-inverting input, resulting in cancellation*

Because of the noise-cancelling action, a twisted pair cable need not (in theory) be screened. This type of cable is commonly referred to as *unshielded twisted pair* (UTP). However, in cases where large amounts of RFI are anticipated, a screen is recommended as it provides added protection against induced noise. This cable type is known as *shielded twisted pair* (STP). Note that because of the action of the twisted pair, mains hum introduced into the pair via ground loops flowing through the screen is cancelled in the same manner as any other noise signal, and thus the inclusion of the screen poses no problems in this area.

There are two practical issues that must be addressed when employing twisted pair cable for CCTV signal transmission. First, it is not possible to fit a BNC connector onto a twisted pair cable. Second, twisted pair cable has an impedance in the order of 100–150 Ω, making it incompatible with the 75 Ω impedance of all CCTV equipment. To overcome these issues, the signal output from the BNC connector on the camera is fed immediately to a twisted pair transmitter, which both isolates the twisted pair from earth, and places the video signal across the two conductors. The transmitter also provides impedance matching between the 75 Ω co-axial cable and the 100–150 Ω twisted pair cable. At the other end of the twisted pair, an impedance-matching receiver places the video signal back onto a 75 Ω co-axial output to facilitate connection to the following item of CCTV equipment. The equipment arrangement for a twisted pair installation is shown in Figure 2.17.

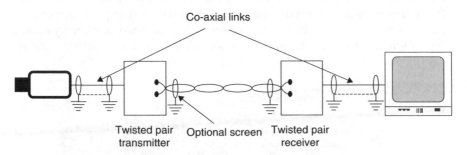

Co-axial links

Twisted pair transmitter Optional screen Twisted pair receiver

Figure 2.17 *Twisted pair transmission arrangement*

Because the twisted pair receiver contains an active electronic circuit (i.e., the op-amp signal processor), a power supply is required. For the twisted pair transmitter, both passive and active devices are available. Clearly, the passive transmitters do not require a power supply; however, the active devices generally offer much greater cable distances.

There is no reason why twisted pair and co-axial cabling cannot co-exist in a CCTV installation. Shorter cable runs may be made using co-axial cable, and longer runs, where signal loss and ground loops might prove problematic, may be made using twisted pair cable. In some CCTV telemetry control systems, it is necessary to run a twisted pair cable alongside the co-axial video signal cable to carry the telemetry data to the pan/tilt/zoom (PTZ) units or dome assemblies. Where the installer has chosen to use twisted pair for both video and telemetry signals, most transmission systems will permit both the video and data signals to be transmitted along a single, four-core cable containing two separate twisted pairs, without interference or cross-talk.

The primary advantage of employing twisted pair video signal transmission in CCTV (compared to co-axial cable) is the much longer distances possible owing to the lower signal attenuation in the cable. Where a high-grade cable (i.e., CAT 2 or better) is used along with active transmitters and receivers, a distance of 1000 m for a colour signal transmission is easily possible, and manufacturers frequently quote figures in excess of 2000 m for a monochrome signal.

The main drawback with using twisted pair is the need for the transmitter and receiver at each end of every cable run, which inevitably increases the cost of the installation. Multiple channel *receiver* units are available which reduce both the installation cost and the number of separate boxes scattered behind the control console.

A cost-saving alternative may be to employ equipment at the control room end that offers direct twisted pair inputs as well as 75 Ω co-axial connection. A typical example is shown in Figure 2.18. To make effective use of this feature, the installer may still employ transmission equipment at each camera, or alternatively they may employ cameras that offer a direct twisted pair output. The advantages are clear: cost saving in receiver (and possibly transmission) hardware, reduced losses (there are always some losses when converting from one transmission format to another), and no need to locate the receive (and possibly transmit) units.

Figure 2.18 *A DVR/MUX which offers direct twisted pair video signal input. (Photo courtesy of Tecton Ltd)*

Structured cabling

In Chapter 11 we will look at computer networking and its applications to CCTV. Here in this chapter we will consider the structured cable systems employed in *modern* networks; i.e., CAT 5, CAT 5e and CAT 6 (at the time of writing CAT 7 is still under discussion). The EIA/TIA (Electronics Industries Association/Telecommunications Industry Association) have provided standards for the categories of twisted pair cable systems for commercial buildings. This is the *EIA/TIA-568 Standard*, which was adopted by the American National Standards Institute (ANSI), although in Europe these same standards can be found in BSEN 50173. The 568 standard can be divided into two: 568A and 568B. However, for the purposes of this text, the main difference can be taken to be the accommodation of Tx/Rx crossover connection in the RJ 45 pinout wiring convention (see Table 2.3). The standards are constantly being revised and updated to keep pace with the rapidly advancing technology that is associated with data transmission, which is apparent when one looks at the number of revisions to date there are for the 568 standards. An outline of categories 1 to 7 can be seen in Table 2.2. The specifications for each category encompass not just the cable but the complete data transmission system including data transmission rates, system topologies, cable specifications, maximum cable and patch lead lengths, termination impedance, hardware (for example, network cards, hubs, etc.) specifications and installation practice.

Table 2.2 *Maximum test frequency and maximum data rate for categories 1 to 7 from the IEA/TIA-568 Standard*

Category	Max test frequency	Max data rate (bps)
1	none	Voice only – not used for data
2	1 MHz	4 M
3	16 MHz	10 M
4	20 MHz	16 M
5	100 MHz	100 M
5e	100 MHz	1 G
6	250 MHz	1 G
7	600 MHz	1 G (at time of writing, a 10 Gbps option is being discussed)

In the networking world, the pressure is on to achieve ever-increasing data transmission rates and this has seen the industry progress from the very early CAT 1 (1 Mbit per second (bps)) to the current CAT 6 (1–10 Gbps). But what determines the maximum data rate through a cable? Well primarily, the cable bandwidth. Referring again to Figures 2.1 and 2.2, we know that square-wave signals are effectively filtered by the capacitive and inductive effects of the transmission cable. Therefore, cable manufacturers are constantly being challenged to develop cables having properties that offer minimum filter action and therefore maximum bandwidth at high transmission frequencies.

CAT 5 and CAT 6 networks all employ UTP cables, designed to have an impedance of 100 Ω. CAT 7 will employ STP, where each pair will be individually screened, with an overall outer screen/shield. The cable specification for the original CAT 5 offered a vast improvement in network bandwidth over the earlier CAT 3 that it was to supersede. We see from Table 2.2 that CAT 3 offered a maximum bit rate of 10 Mbits per second whereas, when CAT 5 was agreed in October 1995, speeds of up to 100 Mbits per second became available. As each new specification is introduced, to date, it has always been conditional that the new networks are backwards-compatible.

The CAT 5 standard became obsolete in May 2001 when the EIA/TIA agreed new standards for an enhanced CAT 5 cable – CAT 5e. This cable type is simply a more refined version of the original, offering more reliable data transmission at higher data rates. Both CAT 5 and CAT 5e have been quoted as being suitable for gigabit Ethernet, offering speeds of up to 1000 Mbps. However, *reliable* data transmission at such high rates only really became possible with the introduction of CAT 5e, and in reality many networks continued to operate at 100 Mbps.

In June 2002 the EIA/TIA agreed CAT 6. In spite of the greatly improved bandwidth figure of 200 MHz (which is what is normally quoted the figure of 250 MHz in Table 2.2 is the *maximum* test figure) at first glance it is difficult to see any significant difference between the new standard and CAT 5e, both being capable of 1 Gbps data transmission. However, a closer look at the specification reveals that the crosstalk and noise figures for CAT 6 are far superior to those for CAT 5e. Furthermore, for CAT 6, it is not only the cable specifications that have been refined. The specifications for connectors, patch cords and network device input/output chip sets are also included in the specification and are far more stringent. The vastly superior CAT 6 specification means a much more reliable data transmission when compared with any of the earlier standards. Finally, CAT 5/5e networks only use two of the four available cable pairs (leaving the other two free for other applications such as telephone connection, or even a second network), whereas CAT 6 utilizes all four pairs (see Table 2.3).

For Ethernet communications, you will often see terms such as *100BaseT* being used. The first figure (100) indicates the data rate – in this example 100 Mbits per second. 'Base' means that baseband signalling is employed, and 'T' indicates that the system uses twisted pair cable.

Connection to CAT 3, CAT 5 and CAT 6 networks is via the *RJ 45 (Registered Jack) plug/socket* (see Figure 2.19). This requirement is a part of the backwards compatibility which was mentioned earlier, but it must be remembered that CAT 6 networks require RJ 45 connectors which meet the higher CAT 6 specification. This is because a network will always function at the rate of the slowest device in the data chain. Therefore if, for example, a CAT 6 network were to include CAT 5 connectors or a CAT 5 hub, that network would only be capable of data transmission at the old 1995 CAT 5 standard. This point has important ramifications for the CCTV engineer who has been tasked to install IP cameras onto an existing network. The customer may be confident that they have a high-performance CAT 6 network which will easily handle the added load imposed by the IP cameras. However, if that customer has unwittingly included CAT 5 or CAT 5e components on their network, the CCTV system may well not perform to standard. Cameras may go off line, picture frames may be continuously lost, or other effects may be noticed on the network such as printers running very slow, email taking much longer, etc. The poor CCTV engineer will often get the blame for 'breaking the network', but in truth the problem was potentially always there – it just required the added load on the network to bring it to the surface.

The four twisted cable pairs are identified by their colours – brown, blue, orange and green – and it is important during installation that the pairs are maintained throughout. Using, say, one green wire and one blue wire as a pair for data transmission would render the noise-cancelling and crossover-cancelling properties of the cable ineffective, and it is unlikely that the network would function. Table 2.3 shows the EIA/TIA wiring conventions for 568A and 568B, for both CAT 5 and CAT 6 installations. Note that for CAT 6, pins 1, 2, 3 and 6 are the same convention as for CAT 3/5, enabling backwards compatibility.

Figure 2.19 *RJ 45 plug and socket connectors. Note that although physically the same, connectors for CAT 3, CAT 5, CAT 5e and CAT 6 differ in their specifications. For example, fitting CAT 5 sockets into a CAT 5e network will reduce the performance bandwidth to that of CAT 5*

Table 2.3 *EIA/TIA-568 Standard wiring conventions*

RJ 45 connector pin	Wire colour (T568A)	Wire colour (T568B)	CAT 3 CAT 5 CAT 5e	CAT 6	PoE connections
1	white/ green	white/ orange	Transmit+	Tx_D1+	Mode A+
2	green	orange	Transmit−	Tx_D1−	Mode A+
3	white/ orange	white/ green	Receive+	Rx_D2+	Mode A−
4	blue	blue	Unused	Bl_D3+	Mode B+
5	white/ blue	white/ blue	Unused	Bl_D3−	Mode B+
6	orange	green	Receive−	Rx_D2−	Mode A−
7	white/ blue	white/ blue	Unused	Bl_D4+	Mode B−
8	brown	brown	Unused	Bl_D4−	Mode B−

In any transmission system, it is necessary to connect the transmit link (Tx) to the receive link (Rx). In an Ethernet network, devices are normally connected to each other via hubs, switches, bridges, etc., and crossover of the Tx and Rx links takes place at these points. Where it is necessary to directly connect, say, two PCs, a *crossover patch cable* must be used, otherwise communication will not be possible because Tx will connect to Tx and Rx to Rx. Practical connection arrangements for both straight and crossover cables are shown in Figure 2.20.

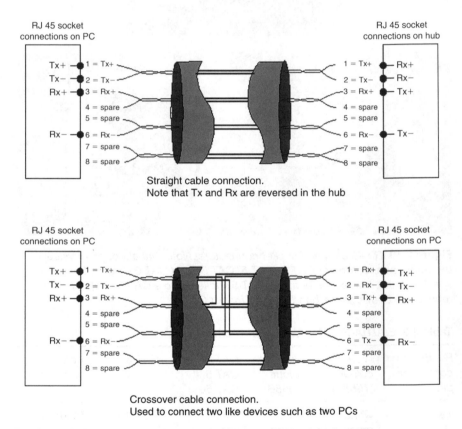

Straight cable connection.
Note that Tx and Rx are reversed in the hub

Crossover cable connection.
Used to connect two like devices such as two PCs

Figure 2.20 *Wiring conventions for EIA/TIA cables and sockets. Note that the colours used will differ for 568A and 568B network connectors (see Table 2.3)*

Another point to be aware of is that CAT 3 and CAT 5 cables physically look very similar, so it is always a good idea to verify that the correct cable has been used for the network installation. Also look out for excessive bending, stretching or crushing of the cable as all of these will alter the cable properties and can result in excessive data errors and subsequent system failure.

Power over Ethernet

A recent development in the field of networking is the *IEEE 802.3af Power over Ethernet* (PoE) standard which was adopted in June 2003. The principle is very simple: use the existing Ethernet cables to carry power to network devices, rather than run separate cables or go to the expense of installing 230 Va.c. supply outlets at the location of every Ethernet device. The system is designed to accommodate power transmission using either the spare pairs in a CAT 3/CAT 5 Ethernet cables, or the actual data lines.

With PoE, the available power on any network segment is 48 Vd.c. at 350 mA maximum, or 16.8 W. The limitation of 350 mA is necessary to prevent overload of the relatively small cross-sectional area network cables. A PoE power supply is required, and this unit must incorporate over-current and over-voltage protection. The protection is two-way: it ensures that network devices are not damaged in the event of a power supply failure, and it also ensures that the power supply itself is not damaged due to a faulty network device or cable short circuit.

The power supply is normally located in the main wiring cabinet containing the network routers/switches. For existing installations where PoE is to be included, a PoE midspan hub is added, and the network segments that are to employ PoE are patched to the existing switch. Power would now be available to all network devices on these segments. The arrangement is illustrated in Figure 2.21. For new installations, Ethernet switches incorporating a PoE power supply are becoming increasingly available, negating the need for a midspan hub.

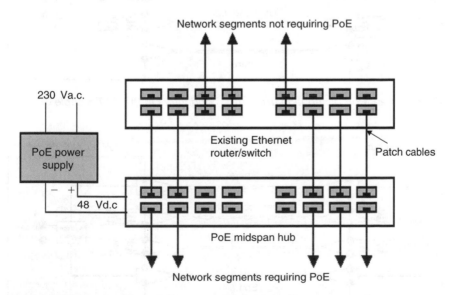

Figure 2.21 *Connecting a Power over Ethernet (PoE) power supply to an existing network using a midspan hub*

Figure 2.22a shows how PoE is delivered using the data cables. Power is injected via a centre-tapped transformer, which means that, from a d.c. point of view, both conductors are in parallel. However, from an a.c. (data signal) point of view, there is no d.c. offset between each conductor. The centre-tapped pickup transformer in the network device is often referred to as the *picker* or *splitter*. This extracts the 48 Vd.c. supply from the two conductors and also, via transformer action, recovers the data signal.

Figure 2.22b shows the circuit arrangement where the spare conductors in a CAT 3 or CAT 5 cable are being used for PoE. Note that both conductors in each pair are being used in order to increase the cross-sectional area and so reduce power loss in the cable. A voltage regulator or d.c.-to-d.c. converter in the network device is usually required to reduce the supply voltage from 48 V to, usually, 12 V or 5 V.

The standard colour code for PoE wiring will obviously have to follow the same convention as the data wiring, because they are in effect the same thing. However, to ease identification, the colours are included in Table 2.3, where Mode A denotes the colours used when the data lines are to be employed, and Mode B denotes the colours for when PoE is to be delivered using the spare conductors.

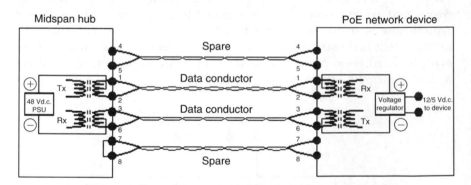

Figure 2.22a *PoE distribution using data conductors in the CAT 5 cable*

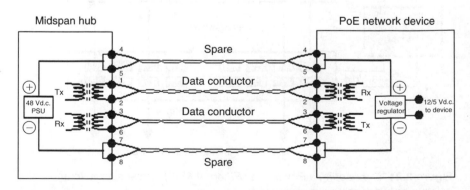

Figure 2.22b *PoE distribution using spare conductors in the CAT 5 cable*

The question must now be asked: what happens if 48 Vd.c. is applied to a network device that does not accommodate PoE? The answer is simple: the device network input circuits will be destroyed! Clearly this issue had to be taken into consideration when the IEEE 802.3af standard was being developed, and a plug and play solution had to be found because there are many non-technical persons who routinely connect network devices such as laptops, etc., to the nearest available network point. Every PoE-enabled device must have a 25 kΩ resistor connected across its input, causing it to draw a current of 2 mA. Whenever a device is connected to a PoE-enabled network link, the midspan hub does not immediately apply the full 48 Vd.c. Instead, it applies a lower test voltage and measures the load current. In effect it is probing for the 25 kΩ resistor. If it detects a resistor, the hub will apply the full 48 Vd.c., otherwise the probe voltage is removed and the network link is treated as an ordinary data link. This method both prevents accidental damage to non-PoE devices, and permits devices on a network segment to be exchanged.

The implications of PoE for CCTV are obvious: network IP cameras are becoming increasingly popular for a number of reasons, including the ease of wiring (the other reasons will be considered in Chapter 11). However, until the introduction of PoE, a d.c. supply voltage still had to be provided for each camera. PoE-enabled IP cameras are becoming increasingly available, and as they do so, the industry will lean ever more towards them for many applications.

Ribbon cable

Also known as 'flat twin' cable, this has two parallel conductors and functions on the balanced transmission principle we have just been discussing. Because the conductors are not twisted it cannot be guaranteed that they will both be subjected to identical amounts of noise energy, although in practice over short distances this will usually be the case. A typical ribbon cable construction is shown in Figure 2.23.

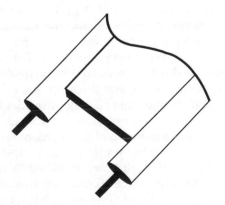

Figure 2.23 *Ribbon cable construction*

This type of cable is useful for interconnection between equipment in a control room, especially for desk-mounted units where a larger, more rigid cable type can prove cumbersome.

Fibre-optic cable

Fibre-optic signal transmission was largely pioneered by the telecommunications industry, and for many years it remained very much within that industry. Perhaps this was because of the specialized skills and equipment required to install fibre-optic cables, particularly in relation to joining (splicing) and terminating. Or perhaps it was due to the relative higher cost of the cable compared with co-axial or other copper transmission medium. Whatever the reason, the CCTV industry was slow to pick up on this very effective method of sending CCTV signals over any distance.

Because the signal travelling through a fibre-optic cable is in the form of light, the medium is not prone to any of the problems associated with copper transmission systems such as RFI, EMI, lightning, etc. Yet fibre-optic transmission has a much wider bandwidth and much lower signal attenuation figures, which means that signals can be sent over far greater distances without the need for any line-correction equipment. Fibre-optic cable also provides complete electrical isolation between equipment so there is never any chance of a ground loop, and from a security point of view it is almost impossible to tap into without it being obvious at the receiving end. The construction of both single fibre and fibre bundle is illustrated in Figure 2.24.

One of the greatest problems associated with signal transmission through optical cable is that of *modal distortion*, which is caused by the light energy finding a number of different paths through the cable. Because the path lengths are not all the same, a single light pulse with a duration of, say, 1 ns applied at the input arrives at the output over a period of around 2 ns. In other words, the information becomes distorted. The longer the cable run, the more acute the problem.

The degree of modal distortion per unit length is determined by the construction of the fibre-optic material, and there are a number of cable types available, each having differing characteristics. In order to minimize modal distortion specially engineered cable must be used; however, the manufacturing of these cables is very expensive. For CCTV systems the cable runs are relatively short (compared with something like a transatlantic undersea telephone cable!), and therefore the effects of modal distortion are minimal and cheaper cable designs are adequate.

Three forms of fibre-optic transmission are illustrated in Figure 2.25. *Monomode cable* is the most expensive of the three, owing to its very small core diameter (typically 5 μm) but it offers the greatest transmission distances with minimal distortion. *Step index multimode cable* employs two different materials in the core, each having a different refractive index. This results in multiple light paths of differing path lengths, which are added

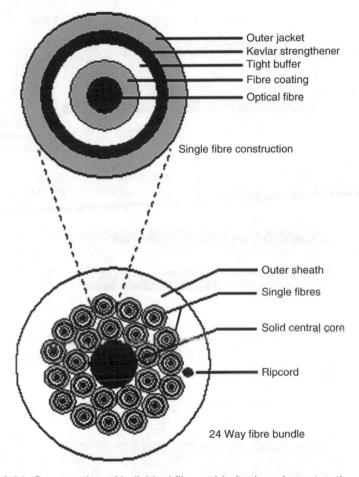

Single fibre construction

Outer jacket
Kevlar strengthener
Tight buffer
Fibre coating
Optical fibre

Outer sheath
Single fibres
Solid central core
Ripcord
24 Way fibre bundle

Figure 2.24 *Construction of individual fibre cable (top), and construction of a fibre bundle (bottom)*

at the output to produce a resultant signal. This cable is relatively inexpensive; however, it offers the greatest degree of signal distortion. *Graded index multimode cable* also employs two different core materials; however, in this case they are diffused into each other, resulting in a multitude of refractive indexes which causes the light to effectively 'bounce' from side to side as it passes along the cable. Graded index (GI) cable offers much better distortion figures than step index cable and is much less expensive than monomode cable, making it the prime choice for CCTV applications. For both step and graded index cables the core diameter is much larger than for monomode cable (typically 50 μm).

With this particular type of fibre-optic cable, runs can easily reach 2 km and, depending on the frequency of the light source, may extend much further without the need of any signal booster equipment.

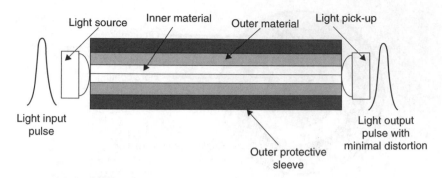

Light source   Inner material   Outer material   Light pick-up

Light input
pulse

Light output
pulse with
minimal distortion

Outer protective
sleeve

Figure 2.25a *Monomode cable. Light travels in a straight line through the inner material*

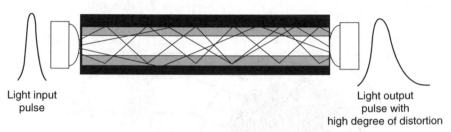

Light input
pulse

Light output
pulse with
high degree of distortion

Figure 2.25b *Step index multimode cable. Refraction between two different materials results in multiple light paths*

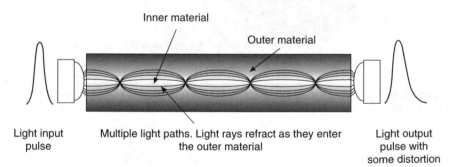

Inner material

Outer material

Light input
pulse

Multiple light paths. Light rays refract as they enter
the outer material

Light output
pulse with
some distortion

Figure 2.25c *Graded index multimode cable. The constantly changing refractive index results in numerous light paths*

The light source may be either a light-emitting diode (LED) or a laser diode. Of the two, the LED is far less expensive, but it has a much lower operating frequency. Having said this, an LED can respond at frequencies of up to 100 MHz, which is more than adequate for CCTV applications. The frequency of light emitted by the LED/laser diode varies between devices; however, some frequencies suffer less attenuation than others therefore offering much longer transmission distances, a factor which manufacturers cannot afford to ignore. The video signal is applied to the light source, which translates the variations in signal voltage into brightness variations.

The light pick-up is a *photodiode,* which is a device that converts light levels into corresponding voltage levels. This forms part of a receiver unit containing the necessary electronic signal processing to produce a 1 Vpp standard video signal on a 75 Ω impedance co-axial output.

As discussed earlier, perhaps the main reason why optical fibre cables are not used more widely on smaller CCTV installations is the cost of installation – and servicing when things go wrong. The cables themselves are expensive; however, on top of this is the cost of employing specialist contractors to perform the critical operations of cutting and splicing the cables, and terminating the ends. Cutting and splicing is critical because any microscopic flaws or inaccuracies in splices will result in the loss of some of the light modes as they deflect into the cladding. An alignment error caused by contamination in the alignment jig is illustrated in Figure 2.26. Cutting the cable requires special tools because, as shown in Figure 2.24, most cables have a Kevlar strengthener running through them and simply chopping through this with a pair of side cutters would result in damage to the optical fibre core.

There are a number of methods for splicing (joining) optical fibre cables, all of them requiring specialized and expensive equipment – and a degree of training. Mechanical splicing involves aligning the two cables in a jig before being either clamped or fixed using an adhesive. Before the two ends can be joined, they must be cut accurately and then polished so that they meet at precisely 90° in order to prevent losses. Alignment is also critical because, when aligning cores with a diameter of just 5 μm, an alignment error of anything more than this will result in a 100% light loss. Another common method for splicing optical fibre cables is fusion splicing, where the two ends are melted together in a process similar to electrical arc welding. This produces a far more accurate splice than mechanical methods and removes the need for such accurate cutting and polishing.

There are a number of ways of multiplexing CCTV analogue video signals so that more than one can be transmitted through a single fibre-optic cable. The actual multiplexing techniques are complex and beyond the scope of this textbook; however, it is useful for the CCTV installer who is considering fibre-optic video transmission to be aware that these multiplexers are available. The alternative to using a multiplexer is to use multi-core fibre-optic cables, but these may prove to be a more expensive option.

It is important to point out the safety issues surrounding fibre-optic cable. Although the cable is quite substantial, the optical fibre itself is thinner than a human hair and particles of fibre can present an almost invisible health hazard to engineers. Small fragments can easily puncture the skin and break off, causing irritation and even infection which can be difficult to deal with medically. Furthermore, broken particles can be inhaled under the right conditions and could remain lodged in the lung indefinitely. Even for engineers who do not deal with fibre-optic cable directly, where their work is following fibre-optic installation, they should check that the area has been properly cleaned, and should avoid allowing their skin to come into direct

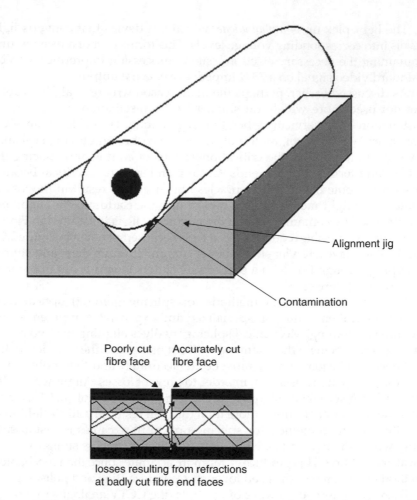

Alignment jig

Contamination

Poorly cut
fibre face

Accurately cut
fibre face

losses resulting from refractions
at badly cut fibre end faces

Figure 2.26 *Alignment loss due to microscopic contamination in the alignment jig*

contact with the floor or other surface areas where they suspect that fibres may still be present.

Infrared beam

In essence this is a variation of fibre-optic signal transmission. An infrared light source is modulated by a video signal, and the light is focused by an optical assembly onto a receiver unit which may be a kilometre or more distant. Both the transmitter and receiver must be able to 'see' each other in a straight line of sight.

As with fibre-optic, there are two types of light source: LED and laser diode. Comparing the two, the LED transmitter is far less expensive, and it produces a wide, diverging beam in the order of 10–20°. However, the diverging beam limits the range to, generally, a few hundred metres. Naturally the limited range can be a problem; however, the wide beam makes for fairly simple system alignment, and the structures onto which the transmitter and receiver units are mounted do not have to be completely stable. The laser diode transmitter, on the other hand, produces a very tight beam of light (around 0.2°) which can travel much greater distances. However, the tight beam requires very accurate alignment, and equipment really needs to be mounted onto solid structures such as buildings to avoid signal loss caused by movement. Also, beware of the effects of direct sunlight, which can cause movement of the metal mountings, throwing the light beam off target.

When locating these units it must be remembered that any break in the light path will result in immediate loss of signal, and although infrared light can penetrate fog and rain to some extent, the range of the equipment will be reduced, and severe weather conditions may result in a loss of signal. Therefore beware of operating these at their specified limits: always allow for some amount of signal loss. Also be aware of other changes which may occur such as leaves appearing on trees during spring, trees that may grow after the equipment has been installed, structures which might be erected, etc.

Infrared links can be very useful for bridging gaps in a CCTV system which would otherwise prove difficult to deal with. For example, it might be decided that an additional camera is required in a town centre system, but its sighting would mean a lot of expensive civil work and much inconvenience to traffic. In these circumstances an infrared link could prove to be a much more cost-effective and desirable solution. A typical application is shown in Figure 2.27.

The equipment usually resembles a pair of small camera housings, and a good quality product will include many of the essential 'extras' which come with camera housings such as wash/wipe, heater, and in some cases an optional fan to cool the equipment during the summer – remembering that extreme heat could cause temporary misalignment of the optics.

Microwave link

The term 'microwave' refers to a band of frequencies in the radio spectrum which extends from 3 GHz to 30 GHz (G = giga, which is 10^9, or 1 000 000 000). These frequencies are above the domestic UHF television channel frequencies and incorporate many of the domestic satellite TV transmissions and mobile phone channels.

In order to prevent interference between signals the air waves are regulated, and no-one is permitted to operate a radio signal transmitter without proper authority, unless it is on one of the frequency bands that have been allocated for free, unlicensed use, and even then the equipment must comply with certain regulations regarding its transmission power and

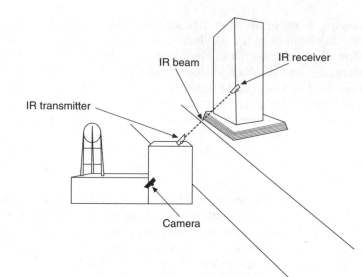

Figure 2.27 *A wide beam infrared link across a main road*

bandwidth. A typical example of one of these 'free for all' channels is the CB radio band. These restrictions mean that CCTV signal transmitting equipment must operate within specified bands, and must not exceed certain power output levels. In the UK, microwave CCTV equipment operates on frequencies located around either 3 GHz or 10 GHz.

Directional (dish) aerials are normally used (Figure 2.28). There are two advantages in this: firstly the transmission range is greatly increased if the power is channelled in one direction, and secondly the signal is more secure against interception by someone operating receiving equipment tuned to the same frequency. However, dish aerials require careful alignment, as just a few degrees of error can result in a loss of signal. This also means that the dishes must be stable. Note that if the dish alignment is only approximate, the signal may produce a perfect picture in good weather conditions; however, as soon as rain or snow settles on the dishes the signal attenuation causes the picture to degrade, with the familiar 'sparkles' appearing. The technology is identical to satellite TV, and this phenomenon is familiar to anyone who has a misaligned or incorrectly sized dish on their home!

Microwave energy does not penetrate solid objects, and thus there must be a clear line of sight between the transmitter and the receiver. This factor tends to limit the use of microwave to short range, unless it is possible to locate the equipment at both ends of the system on top of high structures. Alternatively, the signal would propagate well across flat expanses although there are not many of these in the UK!

In a CCTV application, the transmitter is located close to the camera, connected by a co-axial link. The receiver is located somewhere within line of sight, but in a place where the signal can be sent via cable to the

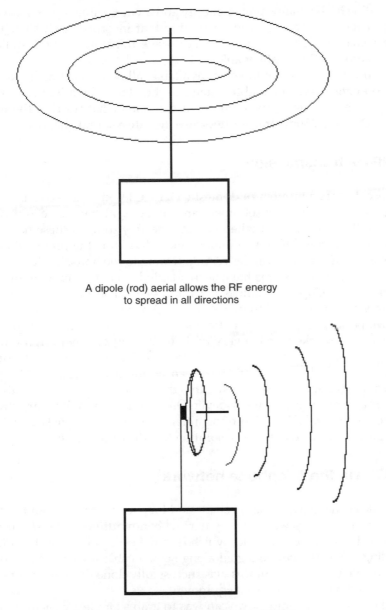

A dipole (rod) aerial allows the RF energy
to spread in all directions

A dish aerial directs the RF energy in one direction,
increasing the transmission range

Figure 2.28 *Comparison of non-directional and directional radio transmission*

control room, ideally as close to the control room as possible, or perhaps at an optical fibre collector point.

Microwave-linked camera installations are proving to be increasingly popular where cameras need to be located in remote areas, and the cost of providing both a mains power supply and a video cable is prohibitive. By employing a camera/housing assembly that incorporates a solar panel and rechargeable battery plus a microwave transmitter, the installation often becomes financially viable.

Two-way microwave links are available, allowing for transmission of video in one direction, and telemetry in the other. The advantage of this can be seen; however, the equipment cost somewhat restricts the use of this technology. Some transmitters also include a sound channel.

UHF RF transmission

In recent years a number of domestic DIY CCTV kits have become available but many DIY householders are unhappy at having to run cables around their homes, and when you see the way some of these have been installed it is not difficult to understand why! Thus the proposition of a wireless CCTV system can be very appealing to the householder.

These wireless systems have been adapted from a technology that was originally developed for domestic TV and VCRs. The idea is to connect the composite video output from a VCR/DVD player into a local UHF transmitter. Because UHF will penetrate solid objects to a limited degree, the signal can be picked up in every room in the house. However, it must be noted that in many cases the signal may also be picked up by the neighbours. This same equipment has been adopted for local CCTV signal transmission, but it must be noted that the range is very poor, and there is no security of the signal whatsoever. On the other hand there are circumstances where a UHF video transmitter designed for domestic VCR use can prove to be a cost-effective solution to a difficult problem.

CCTV via the telephone network

This idea is not new – it has been around for a few decades – but the problem for many years was the very narrow bandwidth of the *PSTN* (public switched telephone network), which is in the order of 4 kHz. The idea of sending a 5.5 MHz video signal along such a cable might at first appear impossible to achieve, but this was successfully done for many years with specially developed *slow scan* equipment.

The principle behind slow scan was to grab a single TV picture frame, digitize it, send it along the PSTN line in the form of a dual-tone audible signal, and then decode this signal at the receiving end. The problem was the time that it took to send the digital information for just one frame, which was in the order of 32 seconds, not including dialling time.

Developments in digital video signal compression brought about an increase in the transmission rate, and *fast scan* was born. Reductions in the amount of data, coupled with increased modem speeds for PSTN lines, led to picture transmission rates of up to 1 TV picture frame per second.

On PSTN the bit rate is in the order of 14 kilobits per second (kb/s); however, this figure was dwarfed by *ISDN* (Integrated Services Digital Network). Developed during the 1970s, this was brought in to cope with the rapidly increasing demand on the telephone network. As the name implies, it is a digital system and therefore offers much higher data rates: 64 kb/s. This can be increased by paralleling lines up to offer $2 \times 64 = 128$ kb/s. In the UK, ISDN lines are installed to order by British Telecom, and a customer wishing to use this for CCTV fast scan must be prepared to pay for the installation and use of this facility.

For many years ISDN proved to be a very effective medium for sending CCTV over long distances without the need for complex and expensive dedicated cable installation. However, more recent developments in Ethernet network technology is rendering this form of CCTV transmission redundant. And, of course, Ethernet itself may be transmitted over the telephone network using ADSL, so CCTV via telephone network is still very much alive.

Connectors

The most common co axial cable connectors in CCTV are the BNC (Bayonet–Neil–Concelman), phono (also known as RCA) and SCART (Syndicat des Constructeurs d'Appareils Radio Recepteurs et Televiseurs). Another type of connector that is very effective and robust is the UHF (PL259) type, but this has never really been adopted extensively in CCTV, and tends to be looked upon as something of a nuisance when it does appear because often engineers do not carry replacements, couplers or converters. A range of connectors is shown in Figure 2.29.

Of the other three types, BNC is by far the most robust, and the locking action of the bayonet fitting means that it does not easily come adrift. Furthermore, its construction and method of termination of the screen and core means that it maintains the characteristic impedance of the cable and does not introduce a significant amount of signal attenuation, provided that good quality connectors are used. The best types have a gold-plated inner pin; or gold-plated inner ferrule in the case of the female connector.

BNC connectors for cable mounting are available as crimp, twist-on and solder fitting. Crimp types are by far the most suitable for CCTV applications, as long as the correct crimping tool is used. Flattening the flanges with pliers, wire cutters, hammers, etc., does not produce a low-impedance, robust fixing, and where this practice is carried out installers can expect to have problems such as intermittent signal loss, ghosting, poor contrast, noisy picture, and loss of telemetry, to name just a few. Twist-on types are very quick and simple to fit, and no special tools are required, but care must be taken to make certain that the screen is compressed tight against the metal body. If this is not the case, then a high-impedance connection exists, if not immediately, then perhaps some time afterwards if the copper screen is

Figure 2.29 *Range of connectors. From left to right: BNC crimp, BNC T coupler, BNC straight coupler, two variations of BNC/phono adapter (top), Phono, UHF, SCART.*

exposed to a damp or corrosive atmosphere. Soldered types make a very strong, low-impedance connection, but they can be extremely messy to fit and the operation is not one which an engineer would wish to perform on site, especially whilst working outside, 30′ high in a snowstorm!

As their name implies, phono connectors were not originally intended to be used with video signal cables but rather for connecting between a phonograph (record player!) and the associated radiogram (hi-fi system!). However, some equipment uses phono sockets for input/output connections because it keeps the cost down. Where this is the case the installer has no choice but to either fit a phono plug to the co-axial cable, which can prove very difficult, or fit a BNC to phono adapter, which introduces another set of contacts, thus increasing the signal attenuation. Having said this, where these sockets are used on a monitor or VCR, the chances are that the connecting leads will only be short interconnecting cables between items of equipment, and in this case ready-made phono–phono leads using simple screened cable will generally perform satisfactorily.

The SCART connector was originally designed for domestic TV and video equipment. During the 1980s it became clear that the television receiver of the future was going to have to be able to accept more than one input source. Furthermore, it was ridiculous to employ UHF RF coupling when equipment such as VCRs were capable of sending a much cleaner composite video signal. Thus a number of leading manufacturers worked to develop a multi-pin connector which would facilitate VCRs, a satellite receiver, stereo audio and RGB input and output links, all on separate screened cables contained within a multicore cable. As can be imagined, initially there were a number of variations, but thankfully an international standard was agreed. SCART was never intended for use in industrial applications, but nevertheless they are found on some VCRs and monitors. The assigned pin connections for a SCART connector are given in Figure 2.30.

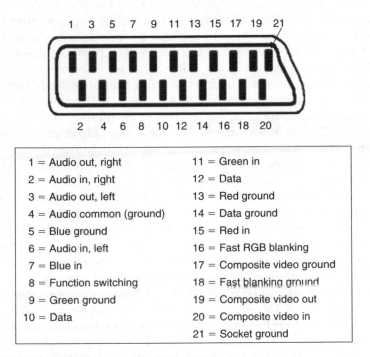

1 = Audio out, right	11 = Green in
2 = Audio in, right	12 = Data
3 = Audio out, left	13 = Red ground
4 = Audio common (ground)	14 = Data ground
5 = Blue ground	15 = Red in
6 = Audio in, left	16 = Fast RGB blanking
7 = Blue in	17 = Composite video ground
8 = Function switching	18 = Fast blanking ground
9 = Green ground	19 = Composite video out
10 = Data	20 = Composite video in
	21 = Socket ground

Figure 2.30 *SCART connector configuration*

Another common connector type is the four-pin miniature Din specially developed for S-VHS applications (Figure 5.14). We shall look at these in more detail in Chapter 5.

When fitting co-axial connectors, care must be taken to ensure that there are no short circuits between the screen and core. A single strand of wire between the two conductors and the signal is lost. Careful and accurate stripping of the outer and inner insulators is the key to avoiding short circuits, and although a sharp pair of wire cutters may be used for this purpose, wire strippers specially designed for RG-59 cable make the task much faster and more reliable.

Cable test equipment

Perhaps the simplest method for testing co-axial cable is to use a multimeter and test for an open circuit between the core and screen when the cable is open at both ends, and 75 Ω when one end is terminated. This, however, will only tell us that there are no short or open circuits in the cable – it tells us nothing about the signal (picture) quality, and the whereabouts of a fault, should one exist.

One method of testing the signal quality capability is to pass a specific test signal along the cable, and measure it at the other end. The test signal could be a sync pulse and black and white test bar pattern derived from a video signal generator. Alternatively a 'pulse and bar' generator may be used. This piece of equipment generates a continuous black, white and grey pattern without any sync pulses or vertical blanking period, and is ideal for signal analysis because of its square waveshape – remembering that a square wave is made up from an infinite number of odd harmonics, so if the cable is introducing any high-frequency attenuation the pulses will appear to have rounded edges (refer to the filtering action illustrated in Figure 2.2). The measuring equipment at the output end of the cable on which the waveform is monitored would be an oscilloscope. (For instructions on how to set up an oscilloscope to observe a video signal, see Chapter 13.)

When the pulse and bar test indicates that a cable is introducing h.f. attenuation, the engineer has to decide what to do about it. If the cable run is only a few tens of metres in length, there is no reason why attenuation should occur, and the cable itself must be suspect, unless it was damaged during installation. Either way, it would have to be replaced. If the cable length is a few hundred metres, then it is clear that some form of correction equipment should be installed; e.g., a launch amplifier. When these have been installed, the pulse and bar generator may be used to check that the signal is correct. In many cases it is necessary to adjust the launch amplifier response to obtain a correct waveshape, and it is not only rounded edges that are a problem – in some cases there is excessive h.f. gain and the waveshape appears to have spikes (called overshoots) on it.

The problem with oscilloscopes is that they are rather bulky to carry around, and although there are some hand-held units available, the LCD display makes them expensive, and it would be something of a luxury for every engineer in a larger company to be equipped with one. However, it is not always necessary to actually see the video signal – the engineer may only wish to know that it is present, and that it is of the correct amplitude. To provide this information without the need to set up an oscilloscope there are a number of video signal strength indicators on the market. A typical example is illustrated in Figure 2.31, where a signal generator is connected to one end of the cable under test, and a strength meter with an LED display is connected to the other. Alternatively, the meter can be substituted with a monitor, and the black and white bar display can be viewed, although with this arrangement the engineer would not know whether or not the amplitude is correct.

Another item of test equipment for use with co-axial cables is the *time domain reflectometer* (TDR). This takes advantage of the reflected waves that occur in a co-axial cable when it is either incorrectly terminated, or is not terminated at all. Both ends of the cable are disconnected, and the TDR is connected to one end. When switched on, the unit sends a succession of short- and long-duration pulses which travel along the cable until they

Figure 2.31 *A typical hand-held test instrument for testing co-axial cables, checking video signal levels, verifying camera output, etc. (Courtesy of ACT Meters Ltd)*

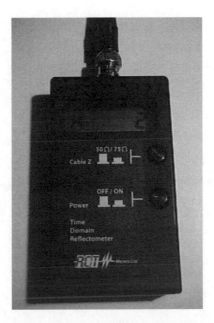

Figure 2.32 *A time domain reflectometer is an extremely useful item of test equipment for installation and service engineers alike. (Courtesy of ACT Meters Ltd)*

come to the unterminated end, whereupon they reflect back down the cable and are picked up by a receiving circuit in the TDR. Because the speed at which the signal travels along a co-axial cable is known (approximately 200 Mm/sec), the TDR is able to calculate the cable length by analysing the time it takes for a pulse to return once it has been sent.

The TDR is used primarily to detect the *position* of a fault in a cable, which is a very useful thing to know when faced with a cable length of a few hundred metres, installed in a building, and all you know from your multimeter is that 'somewhere' along the cable there is a short or open circuit. Upon reaching the fault in the cable the pulses reflect, and the TDR calculates the distance to the fault, usually indicating this on a digital display (see Figure 2.32). The TDR is remarkably accurate, and even over a distance of a few hundred metres it can be accurate to within just two metres.

Apart from detecting cable faults, the TDR can be very useful for determining the length of cable left on a roll, rather than having to reel it all out and measure it. The TDR can also indicate the location of an excessive bend in the cable that is impairing its performance.

An optical TDR is available for use with fibre-optic cables, and like the co-axial version, it is an essential part of a service kit for anyone dealing with fibre-optic cables regularly. Although the optical unit is somewhat more expensive, it can very soon pay for itself by making savings in both time and materials.

3 Light and lighting

Although the majority of the equipment in a CCTV system is in the business of processing electronic images in some form or another, the lens has the task of processing the light that is reflected off the surface of the target. The importance of this task cannot be stressed enough because, if the lens fails to focus a true image onto the camera pick-up, then the rest of the system will have no chance of producing faithful and useful images. Yet before we can look at the operation of the lens, we need to be clear on the nature of the quantity that it is processing, so in this chapter we will look at those aspects of light and lighting that appertain to CCTV system design and operation.

The world's leading physicists are still uncertain about the true nature of light. However, they are generally agreed that it is made up from minute particles called photons. But this photon energy is electromagnetic in nature – just like radio waves – and as such propagates sinusoidally, therefore having a *frequency* and thus *wavelength* (the higher the frequency, the shorter the wavelength). The *frequency of a light wave determines the colour*.

The spectral distribution for visible light, as well as the infrared and ultraviolet regions, is shown in Figure 3.1. As you can see, the human eye

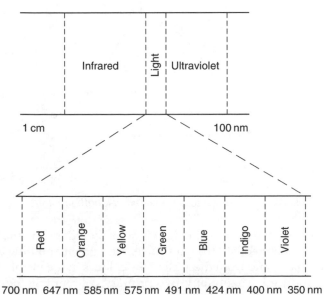

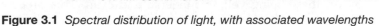

Figure 3.1 *Spectral distribution of light, with associated wavelengths*

begins to respond to frequencies of light energy at wavelengths of around
700 nm (nanometres), which corresponds to the red region, and loses its
response in the blue region at around 350 nm. However, we must remember that the camera pick-up device can have a much wider response, particularly in the infrared region, making it capable of producing a picture in
what would appear to us to be total darkness.

Light and the human eye

The eye reacts to light, converting the incoming electromagnetic radiation
into small electrical signals which are sent to the brain. The brain converts
these signals into an image.

One set of cones responds to the range of frequencies with corresponding
The eye contains four sets of cells; the cells of one set have a cylindrical
structure and are known as rods. The other three sets are conical in shape
and are known as cones. The cones are sensitive to different frequencies of
light, and it is these which enable the eye to differentiate between colours.

One set of cones responds to the range of frequencies with corresponding
wavelengths in the order of 600–700 nm. Upon receiving signals from these
cones the brain acknowledges that it has seen red light. The second set of
cones responds to wavelengths around 500–600 nm, that is, green light,
and the third set responds to 400–500 nm, blue light.

The eye does not respond equally to all frequencies, nor is the response
equal for all people. As Figure 3.2 illustrates, a good eye responds best to
wavelengths of around 550 nm – that is, green light.

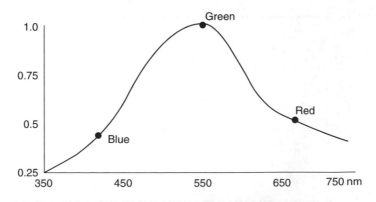

Figure 3.2 *Response of the eye to different wavelengths of light*

What we perceive to be white light is actually red, green and blue light
emanating simultaneously from a source. Red, green and blue are known
as *primary colours*, and from these all of the colours in the visible light spectrum are produced. Mixing any two primary colours together produces
a *secondary colour*, these being yellow, magenta and cyan. This process,
known as *additive mixing*, is illustrated in Figure 3.3.

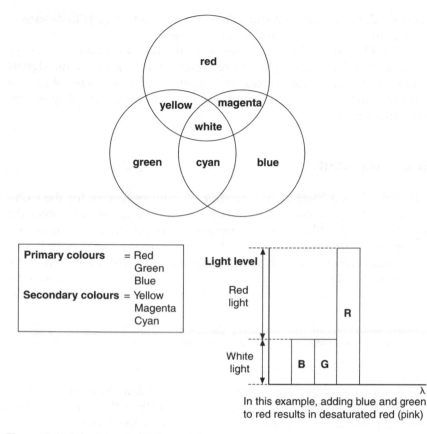

Figure 3.3 *Principle of additive mixing*

We use the term 'colour' in everyday language to distinguish between, say, red and blue. But in practice this term actually encompasses two aspects which describe the colour: the *hue* and the *saturation*. Hue describes the frequency of the light, i.e., red, blue, etc. Saturation describes the intensity or strength of the colour. For example, desaturated red is the correct term for pink, because it defines the hue as being red, but because of the addition of small amounts of green and blue, there is an element of white which desaturates it.

The rods are not frequency (and therefore colour) conscious. These cells may be seen as broadband receivers that respond to the entire visible light spectrum, therefore measuring the intensity of the total amount of incoming radiation. They determine if the scene is very bright, very dark, or somewhere in between. If the eye were to contain only rods we would see everything in black and white.

Summarizing, the rods determine the black and white content of an image, and the cones determine the colour content. The brain constantly processes the information coming from the four sets of cells to determine the brightness, hue and saturation of the image.

In a CCTV system, the camera performs the function of the human eye, converting the incoming light into red, green and blue electronic signals. At the other end of the system, the monitor (or other type of display device) has to take these signals and convert them back into red, green and blue light outputs.

Measuring light

In reality this is a complex science but it is not necessary for the CCTV engineer to have an in-depth understanding of the subject. However, it is important to have a basic understanding of the quantities and units used for light measurement as these are frequently referred to in both system and manufacturers' specifications. The unit of light most commonly encountered is the *lux* (lx), so let's see how this unit is derived and in the process we shall also define a few other units that engineers may well come across.

From the point of view of the eye (and therefore the CCTV camera) there are two sources of light: primary sources such as lamps, display devices, etc.; and secondary sources, which are surfaces or objects from which light is reflected.

The light radiated from a primary source is known as the *luminous intensity*. This is measured in *candelas* (cd), which is the amount of light that is radiated in all directions from a black body that has been heated to a temperature equal to that at which platinum changes from a liquid to a solid state.

The *lumen* (lm) is a measure of *luminous flux,* which is the light contained within an area of one radian of a solid angle. One lumen is equal to a luminous intensity of 1 cd within that given area. If this all sounds a bit much, just imagine one lumen as being 1 cd of light within a given area! Because the light is constantly diverging as it moves away from the source, the intensity, when measured at different points away from the source, will reduce. The divergent light reduces by an amount equal to the square of the distance from the source. Therefore, we can say that:

$$\text{Scene illumination} = \frac{\text{light output}}{\text{distance squared}}$$

This effect, which is known as the *inverse square law*, is illustrated in Figure 3.4.

When one lumen of light falls onto an area of one square metre, the surface intensity will be *one lux* (see Figure 3.5).

Another measure of surface intensity is the *foot-candle (or foot-candela)*. This unit, which is commonly used in the USA, is derived in the same

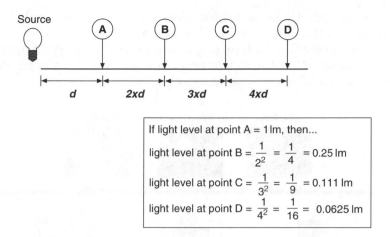

Figure 3.4 *Light illumination reduces by an amount equal to the square of the distance from source*

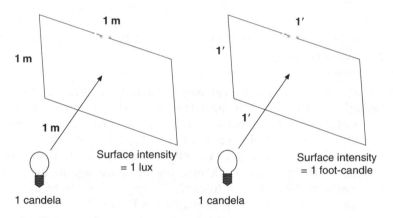

Figure 3.5 *Defining the lux and the foot-candle. As an approximation, we can say that one lux = ten foot-candles*

manner as the lux; however, imperial units of feet are used in place of metres. Conversion between lux and foot-candles is not difficult because three feet approximates to one metre and therefore *as a rule of thumb we can say that one foot-candle = one tenth of one lux or, conversely, one lux = ten foot-candles.*

Illumination refers to the amount of light coming from a secondary surface or object, the unit of measurement being the lux or foot-candle. The material covering or making up a surface will determine the amount of light reflected off that surface, so even if the primary light source has a high intensity, if the reflective quality of a surface is poor, the level of illumination will be low. As a typical example, a white projector screen may reflect 90% of the incident light and will itself serve to illuminate other objects in the area. However, a dark coloured coat may reflect only 5% of the incident light, resulting in low illumination.

Thus we can see that the level of illumination is dependent on both the source lighting and the reflective properties of the surface areas. As a guide, some typical levels of illumination are given in Figure 3.6.

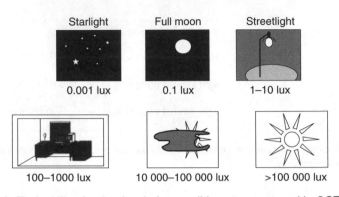

Figure 3.6 *Typical illumination levels for conditions encountered in CCTV installations*

Light characteristics

As far as we are concerned, we can assume that light travels in a straight line through a vacuum. Light also travels in straight lines through a medium, and the speed at which it travels differs according to the density of the medium. For example, when a ray of light passes from air into a solid such as glass, the change in velocity causes the ray to bend. As the ray emerges from the glass back into the air once again, its velocity increases, causing it bend back towards its original angle of incidence. This effect, known as *refraction,* is illustrated in Figure 3.7.

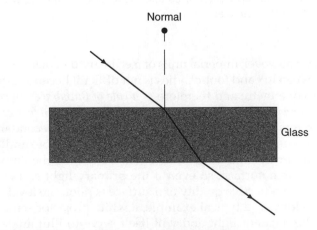

Figure 3.7a *Effect of a solid on the path of a light ray*

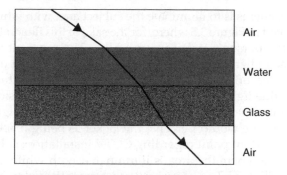

Figure 3.7b *Effect of differing media on a light ray*

The amount by which a ray is refracted depends upon the change of velocity, which in turn is governed by the density of the solid. Each solid has its own *refractive index,* which is found by dividing the velocity of light in a vacuum by the velocity of light within that solid. Air and clear gases are taken as having a refractive index of 1.

However, it is not only the refractive index which governs the angle of refraction. The frequency (i.e., the colour) of the light ray also has a bearing. Experimentation proves that for any given solid, the blue end of the light spectrum will always be refracted more than the red. This is why when we pass white light through a prism, we observe a 'rainbow' effect. Although this effect might look striking in a table ornament, it causes considerable headaches for lens manufacturers, as we shall see in the next chapter.

Artificial lighting

All too frequently lighting is the one factor that the CCTV system designer, installer and end-user have little control over. To guarantee clear, true, colour images from a video camera, the scene should be illuminated with a high level of evenly distributed white light. To illustrate this point, consider the following scenario: a television production company wish to film a scene in a street. Apart from the film crew, sound engineers, production people, electricians, actors, catering van, etc., there will be a dedicated lighting crew who will erect as much lighting and reflective surfaces as they deem necessary to ensure a high-quality picture, free from any shadows. Now the crew have gone away, and all that is left is a CCTV camera covering that same street, and its owner expects it to produce the same quality images in all weather and lighting conditions, with the only form of artificial lighting at night being the sodium street lights. Without the lighting crew, this camera has no chance of meeting this expectation!

Bearing in mind that visible white light actually comprises the three primary colours red, green and blue, it follows that the only way (using current technology) of achieving true colour reproduction with either video or photographic cameras is to illuminate the subject area with white light. This point is illustrated in Figure 3.8 where, for the sake of this illustration, we shall assume that the person's jacket is pure blue. The pigment in the jacket is such that the red and green light energy is absorbed, but the blue light energy is reflected off the surface. Therefore the eye (and thus the brain) discerns only blue light in that part of the image. Now consider the same image if it is illuminated with a pure red light. All of the light would be absorbed, so the eye would perceive the jacket as being black. This leads us to a very important point regarding CCTV installation: colour cameras are only effective where the area is illuminated with white light. *There is no point in installing a colour camera where infrared* (IR) *lighting is to be used* although, as we shall see in Chapter 6, there are some colour cameras that will automatically switch to black and white operation when it is dark or when IR lighting is present.

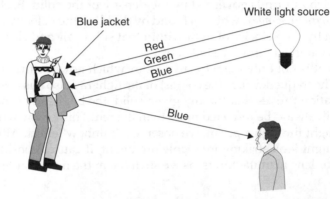
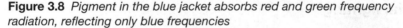

Figure 3.8 *Pigment in the blue jacket absorbs red and green frequency radiation, reflecting only blue frequencies*

There are a number of sources of 'white' light and most are not true white, but nevertheless when used with CCTV equipment many of these will provide adequate illumination. The type of light produced by a source is determined by the *colour temperature*, which is a scientific measure of the wavelengths of light and is stated in degrees Kelvin. For example, the light from a fluorescent tube is known to be about equivalent to that of an overcast day because in both cases their colour temperature is around 6000–6500 K.

The spectral output range for a number of common sources of artificial lighting is given in Figure 3.9. Figure 3.9a shows the overall light output for a number of sources; however, Figure 3.9b gives a much clearer indication of the usefulness of some of these sources. Bearing in mind that monochrome cameras respond well to IR light (unless they have been manufactured with IR filters), then we can see that for monochrome CCTV installations almost any of the common artificial light sources will provide adequate illumination. However, for colour operation we must be more selective.

As a general rule, colour cameras work best with tungsten or tungsten halogen lighting. Although Figure 3.9a would indicate that high-pressure sodium lighting provides suitable coverage of the visible spectrum, in practice this type of lighting produces mainly yellow illumination with only small amounts of blue and red. Fluorescent lighting produces a somewhat irregular output across the visible spectrum, but most cameras function reasonably well under these lighting conditions as long as the tubes are not operating at the 50 Hz (60 Hz) mains frequency, which can result in an undesirable flickering effect on the picture. Note the very narrow response of low-pressure sodium lamps. These produce mainly a yellow/orange light, making them unsuitable for both colour and monochrome cameras; however, until recent years this type of lighting had been used as the main form of street lighting in the UK but much of it is still in use in urban areas.

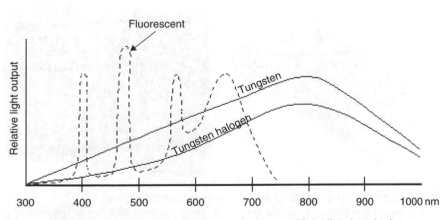

Figure 3.9a *Spectral output range for common sources of artificial lighting*

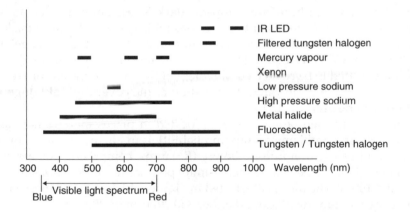

Figure 3.9b *Typical spectral response of common artificial lighting devices*

Another problem associated with gas discharge lamps (i.e., low- and high-pressure sodium) is the time that it takes for them to strike and then reach their optimum light output performance. The time from strike to maximum light output can be a few minutes, and furthermore, if they are turned off, it is not safe to strike them again until they have cooled, which can also take a few minutes. For this reason, these types of lighting are unsuitable for use where switched security lighting is required. On the other hand, although tungsten halogen lamps reach their optimum light output almost immediately, where they are being constantly switched on and off it is generally found that the lamp life is relatively short due to failure of the filament.

Infrared lighting units are usually tungsten halogen lamps with a filter mounted in front. There are two common types in use, one producing illumination around 730 nm, the other having a longer wavelength in the order of 830–860 nm. 730 nm units tend to produce a small amount of visible red light which makes them visible to the naked eye – although the area that they are illuminating appears dark. These units are ideal for use in locations where white lighting is undesirable (perhaps because the local residents do not want their bedrooms illuminated like a theatre stage!) but where overt CCTV is required to serve as a deterrent. Where the CCTV system is required to be covert, employment of IR lamps in the 830 nm region will result in a high infrared illumination of the cover area, although the lamps will be invisible to the eye.

Because infrared lamps are often located in places that are not easily accessible, the problem of lamp unreliability is made more acute. For this reason LED (light-emitting diode) IR lamps comprising an array of infrared LEDs are becoming increasingly popular for applications of up to around 100 W. The idea is illustrated in Figure 3.10. In general these lamps are covert, operating at around either 850–880 nm or 950 nm.

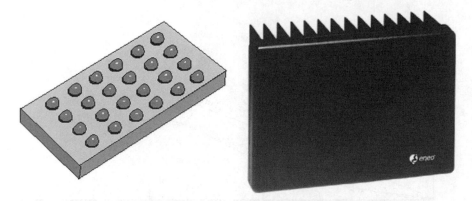

Figure 3.10 *The drawing illustrates the principle of the LED array. The photograph shows a typical lamp unit. Note the heat sink fins at the rear of the lamp which provide convection cooling*

When we think about the intensity of CCTV lighting, too often we are only concerned about the area being too dark. However, too much lighting can also pose problems. Just as the human eye does not cope well with excessive bright light, neither does a CCTV camera. Consider the situation where a fully functional camera (i.e., one mounted on a pan/tilt unit and having full operator control) with artificial lamps is required to cover a large car parking area. In order for the operator to be able to discern activity anywhere in the area it may well be necessary to install two 500 W flood lamps – either white or infrared. However, problems will arise when the camera is tilted downwards to view a target in close proximity because the target will appear completely overexposed. This problem may be overcome by having more than one type of lamp available to the operator, each being independently switched. For example, in the case of the car park, a high-wattage flood lamp could be used when viewing the general area and a low-wattage spot lamp could be switched on when the camera is being used to monitor the area close by.

In general there are three types security lamps available: flood, wide angle, or spot coverage (Figure 3.11). The choice of distribution type is important. A 500 W flood lamp will cover a wide area, but it may not provide sufficient illumination in any particular part of the area. At the other extreme, a 500 W spot lamp will leave large areas with no illumination and might well result in overexposure of the target area. From this we should appreciate that, when installing lighting for CCTV, consideration should be given to such factors as the areas to be covered, the amount of illumination required, and the types of lighting – i.e., white, infrared, overt or covert.

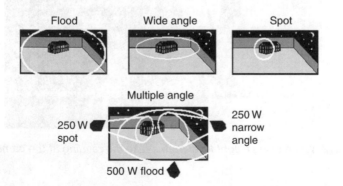

Figure 3.11 *In some situations a mixture of lighting distributions is required to provide adequate illumination*

Consideration should also be given to the angle of the lighting. In general, security lights have to be mounted high up both to ensure their safety and to place them in the proximity of the camera (where the power source and telemetry control will be available). However, if the only lighting source is from above then facial features may be lost due to the heavy shadowing effects around the eyes and mouth. For example, the shadow cast by the nose could well be misinterpreted as a moustache. Also, beware of low-level downlighting used to illuminate some paths and walkways. Whilst the level of illumination can be more than adequate to satisfy the needs of the CCTV camera, this type of lighting tends only to provide illumination of the waist down, therefore making it impossible to identify the persons in the shot.

We have already seen that light levels are measured in lux. The instrument used for taking these measurements is a light level meter, or lux meter. From the point of view of CCTV camera installation, where it is felt necessary to test the light level in an area, the most practical point to take the measurement is at the camera lens because this will give an indication of the actual amount of reflected light entering the lens from the secondary surfaces and objects (see Figure 3.12). When using a light meter, take note that the photo sensor deteriorates with exposure to light. Manufacturers provide a black cover for the sensor and, to ensure accurate readings and preserve the life of the meter, this should only be removed whilst taking a reading.

Figure 3.12 *When using a light level meter, take the reading at the camera location*

4 Lenses

The performance of a CCTV system is very much reliant on the quality and type of lens fitted to the camera. A system will offer poor picture performance when the installer does not specify the correct lens during the initial survey, and 'correct lens' does not simply mean choosing a lens which will offer the correct field of view, although this is one important factor. The quality of the lens, the format size and the spectral response are all-important factors relating to lens performance and thus image quality. For example, there is no point in fitting a lens with an infrared filter when the system is expected to perform in the dark with the assistance of artificial infrared lighting! And this has been known to occur.

In this chapter we shall begin by looking at the principle of operation of optical lenses before moving on to discuss the lenses employed in the CCTV industry.

Lens theory

An optical lens is a device which makes use of the refractive effect on light paths. There are two types of lens: *convex* and *concave*.

A simple convex lens is shown in Figure 4.1a. The light rays entering the lens are refracted, but because the lens surface is curved, the angle of emergence at each point on the lens is different. If the lens is ground accurately, then all of the rays of light will converge at a single point somewhere behind the lens. This point is called the *focal point*.

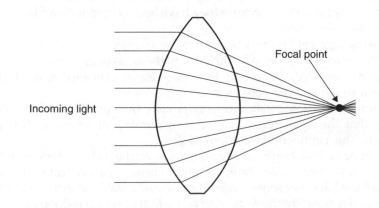

Figure 4.1a *Convex lens. Light converges onto a focal point*

The concave lens is shown in Figure 4.1b. This is known as a diverging lens because the light rays are bent outwards. In this case the focal point is said to be at a point on the entry side of the lens from where the light appears to have originated.

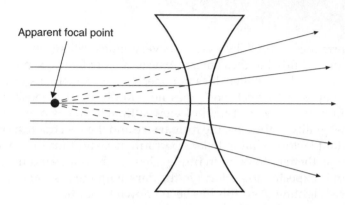

Figure 4.1b *Concave lens. Light diverges from an apparent focal point behind the lens*

One problem encountered with a simple lens is *chromatic aberration*. Different colours of the visible light spectrum have a different refractive index, so for the lens in Figure 4.1a, the red end of the spectrum will not be bent as far inwards as the blue end. This means that there are a number of focal points, i.e., one for each wavelength in the light spectrum, resulting in an image that appears to have a number of coloured haloes around it. In effect, the lens is behaving rather like a prism. The problem of chromatic aberration can be overcome by using a lens assembly comprising a series of converging and diverging lenses, each made from glass with a differing refractive index.

The efficiency of a lens is reduced if the glass is highly reflective because a considerable amount of light simply bounces off the front face. Lenses often have a bluish tinge because they have been coated with a filter material to reduce this effect.

Looking again at Figure 4.1a, we can see another important characteristic of the convex lens. Because the light paths cross over at the focal point, *the image is inverted*. Of course, this inversion is cancelled if the light is passed through another convex lens.

A practical *lens assembly incorporates more than one optical lens*, which means that the light paths may cross a number of times as they pass through. This is illustrated in Figure 4.2.

The image will only appear in correct focus on the pick-up device when all of the lenses within the assembly are at the correct distance from each other. We shall see later that some lenses are made adjustable to offer *zoom effects*; however, *this means that the focus must be re-adjusted*, which is done by moving the position of the lens assembly using an adjustment called the *focus ring*.

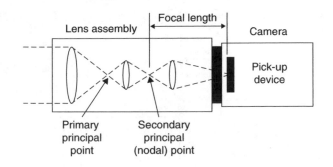

Figure 4.2 *Light paths through a lens assembly*

Lens parameters

Having looked at the basic operating principle of a lens, we can now give a more detailed consideration to the subject. A sound understanding of the meaning of, and the principles behind, the terms *lens format, focal length, angle of view, field of view, aperture, F-number,* and *depth of field,* is essential for anyone who wishes to effectively specify and/or install CCTV systems.

Let's begin with *lens format*. This is directly related to *camera format*, which refers to the size of pick-up device employed by any particular camera. Pick-up devices come in a range of formats (sizes). CCTV cameras originally employed vacuum tube pick-up devices, and the 1″ tube was a common choice, offering reasonable performance at an affordable price. However, tubes have long been superseded by the charge coupled device (CCD). We shall look at CCDs in Chapter 6.

CCDs do not require the same pick-up size as their tube counterparts to produce a picture of matching resolution, which is why ¼″, ⅓″, ½″ and ⅔″ format cameras are common in the CCTV industry. Rapid advancements in CCD chip technology has led to ever-increasing resolution from the smaller chips, making the ⅓″ and ½″ format cameras the most popular choice for general purpose applications. ¼″ CCD chips are very common, but (at the time of writing) cameras employing such chips generally require reasonable lighting levels and are only capable of producing lower-resolution colour images. As the cost of CCD chip production continues to fall, ⅔″ format cameras are becoming increasingly popular; however, relative to their smaller counterparts, these cameras are still more expensive and their use tends to be relegated to high-security CCTV installations, highway traffic monitoring or town centre monitoring systems. 1″ format CCDs are available, offering very high resolution and excellent low light level performance, but at a very much higher cost.

Five formats are shown in Figure 4.3, along with the corresponding horizontal and vertical chip dimensions. The chip size complies with the industry standard ratio (aspect ratio) of 4:3, and these horizontal and vertical dimensions are standard for all pick-up devices. Imperial units are used to denote the camera and lens format to prevent confusion between the lens

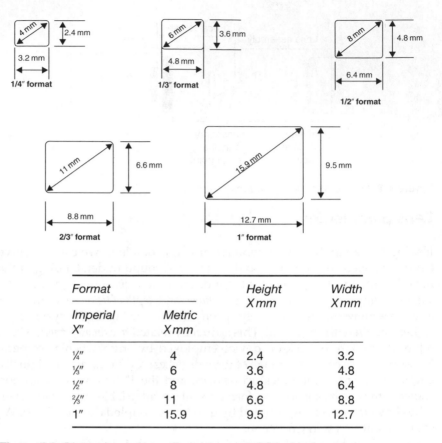

Figure 4.3 *Dimensions of imager size for each CCD chip format*

Format		Height	Width
		X mm	X mm
Imperial X"	Metric X mm		
¼"	4	2.4	3.2
⅓"	6	3.6	4.8
½"	8	4.8	6.4
⅔"	11	6.6	8.8
1"	15.9	9.5	12.7

format and focal length figures. For example, it is much less confusing to talk about a ½" 12 mm lens than it is to refer to a 12 mm 12 mm lens, even though these are more or less the same.

It is important to note that the actual dimensions of the CCD device are smaller than the format size. Take for example the ½" format CCD. You might expect the diagonal dimension to be a ½" (12.5 mm); however, it is only 8 mm. The same rule applies to the other three image devices; their diagonal dimensions are all smaller than their quoted format sizes. The reason for this is that we do not use the light output from the lens at the outer edges of the image, as this is where maximum optical distortion occurs. The relationship between lens and camera is illustrated in Figure 4.4.

Let's now look at what happens when the lens/camera formats have not been correctly matched. In Figure 4.5a the *lens format is smaller than the camera format*, and so the image does not fill the display. When viewed on a monitor, it would appear as if we were looking through a porthole! In Figure 4.5b the *lens format is larger than the camera format*, so in this case the image fills the monitor screen; however, not all of the image produced by the lens is being

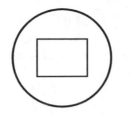

Figure 4.4 *Proportional dimensions of a 1/2" format CCD when compared with a 1/2" diameter circle, illustrating the relationship between lens and camera formats*

(a) A 1/3" format lens fitted
to a 2/3" format camera

(b) A 2/3" format lens fitted
to a 1/3" format camera

Figure 4.5 *Two examples of mixed lens and camera formats, drawn to scale*

used. This is not necessarily a bad thing, because by only using the centre area of the lens there will be minimal optical distortion. However, the operator must be able to view the area specified in the operational requirement.

Moving on to *focal length*, looking again at Figure 4.2 we see that *the focal length is the distance, measured in millimetres, from the secondary principal point to the final focal point*. The secondary principal (or nodal) point is the last occasion where the light paths cross in the assembly, and the camera pick-up device is located at the focal point. For small, wide-angle lenses that have a very short focal length, it would not be possible to fit the lens to the camera because the rear lens assembly would be forced against the pick-up device. In these cases it is necessary to fit additional optical devices to compensate for this, moving the final focal point further away from the rear lens assembly, without altering the effective focal length of the lens.

The focal length has considerable effect on the performance of the lens. The effect of changing the focal length, that is, moving the nodal point, is shown in Figure 4.6, where it can be seen that the *angle of view* changes.

But moving the nodal point not only changes the angle of view; it also affects the *magnification* (M). All people with normal eyesight have more or less the same angle of view (which is approximately 30°) and magnification. Taking this value as a magnification of $M = 1$, it can be shown that a ½" format CCTV camera fitted with an 8 mm lens has a magnification of approximately 1, thus producing an image close to that perceived by the human eye. Table 4.1 shows the focal lengths for other lens formats having a magnification of $M = 1$.

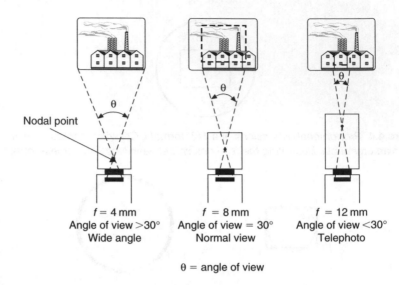

$f = 4\,mm$
Angle of view >30°
Wide angle

$f = 8\,mm$
Angle of view = 30°
Normal view

$f = 12\,mm$
Angle of view <30°
Telephoto

θ = angle of view

Figure 4.6 *Effect of focal length on the angle of view*

Table 4.1 *Lens sizes having a magnification M = 1. In each case the angle of view is approximately 30°*

Format size	1"	⅔"	½"	⅓"	¼"
Focal length	25 mm	16 mm	12 mm	8 mm	6 mm

Increasing the magnification increases the apparent size of an object, and *lenses with a long focal length are known as telephoto lenses* because they have a narrow angle of view, but objects far away appear much larger than they would when viewed with the naked eye. *Lenses with focal lengths that produce an angle of view wider than 30° are known as wide-angle lenses* because they cover a broader area than the eye; however, the magnification is less.

It can be shown that the magnification of a lens is determined by the expression:

$$M = \frac{\text{focal length } (f)}{\text{format size}}$$

where M is the magnification factor, and format size is the size of the lens.

For example, for a ½" format camera fitted with a 12.5 mm (½") format, 25 mm focal length lens, the magnification will be:

$$M = \frac{25}{12.5} = 2 \text{ times}$$

However, if a 100 mm focal length lens is fitted to the same camera, the magnification becomes:

$$M = \frac{100}{12.5} = 8 \text{ times}$$

So we see that *increasing the focal length increases the magnification*. However, the increase in focal length gives a corresponding reduction in the angle of view and thus the image size, known as the *field of view*, is reduced. Putting this another way, increasing the magnification (zooming in) reduces the area that can be viewed.

Field of view is one of the more critical factors in CCTV system specification and design because it determines how large an image will appear on the monitor screen for a given distance from the camera. When we talk about the field of view we need to define whether we are referring to the vertical or horizontal dimension (Figure 4.7).

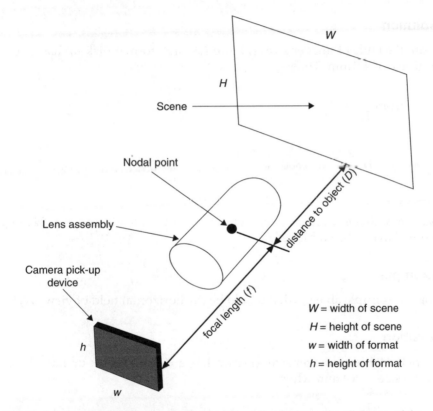

Figure 4.7 *Relationship between the focal length of a lens, the distance of the object from the lens and the field of view*

But field of view is not only affected by the focal length; it is also a function of the format size and the distance from the camera to the object in focus. This relationship is expressed as:

$$\frac{w}{W} = \frac{h}{H} = \frac{f}{D}$$

Where w is the width of format, W is the width of scene or object, h is the height of format, H is the height of scene or object, f is the focal length of lens, and D is the object distance from camera. Note that in some instances the letters 'V' and 'H' are used to denote the *vertical height* and the *horizontal width*. This is particularly the case with some lens calculators (see later in this chapter).

Example

It is required that an object of 2.5 m in height fills the monitor screen when observed at a distance of 15 m. The camera format is ½‴. Calculate the required lens focal length.

Solution

From the table of imager sizes (Figure 4.3), a ½″ format pick-up has a vertical size of 4.8 mm. Thus:

$$\text{from } \frac{h}{H} = \frac{f}{D}$$

$$f = \frac{h}{H} \times D = \frac{4.8}{2500} \times 15\,000$$

$$= 28.8 \text{ mm focal length}$$

Naturally this is an ideal size, and the nearest available size, e.g. 32 mm, would have to be employed.

Example

For the example above, what would be the horizontal field of view (W)?

Solution

From the table of imager sizes (Figure 4.3), a ½″ format pick-up has a horizontal size of 6.4 mm. Thus:

$$\text{from } \frac{w}{W} = \frac{f}{D}$$

$$W = \frac{w \times D}{f} = \frac{6.4 \times 15\,000}{28.8}$$

$$= 3333 \text{ mm or 3.333 m wide}$$

The above examples site the maximum height and width of the field of view, but in some cases we have to calculate the lens size from the point of view of the size of an image on the screen. For example, it may have been decided that in order to be able to recognize a person of average height at a range of 15 m, the image of the person must fill at least 67% (2/3) of the vertical size of monitor screen. Faced with such a situation, the specifying engineer must ensure that the lens fitted to the camera will perform to this specification. This problem may be approached in the following manner:

- Assume that a ⅓" format camera is to be employed, then (from Figure 4.3) $h = 3.6$ mm.
- Assume it has been decided that a person of height 1.6 m will be the minimum height of person we wish to target. For this image to fill 67% of the screen, the full field of view height will be 1.6 m $\times$ 1.5 m = 2.4 m, or *2400 mm = H*.

$D = 15\,000\,mm.$

Thus

$$f = \frac{3.6}{2400} \times 15\,000 = 22.5\,\text{mm}$$

In this example it was simple to find the full field of view height, because it was a convenient number: 2/3. An alternative method of calculating this figure when the percentage screen height is not so straightforward is:

$$H = \frac{100\%}{\text{target height (\%)}} \times \text{target height (metres)}$$

To prove this equation, let's insert the figures from the last example...

$$H = \frac{100\%}{67\%} \times 1.6 = 2.39\,\text{metres}$$

As a general rule, for any format size, *doubling the focal length halves the field of view, and doubles the magnification*. In other words, we only see half as much, but it appears twice as large. This is a useful rule of thumb when working on site.

Calculating the lens size for a particular application is very important. Failure to get this right at the specifying stage can prove expensive in the long run when a customer refuses to pay the bill because the images produced on the screen are nothing like those laid down in the specification, or *Operational Requirement* (OR). Given the cost of some lenses it could prove expensive to go around changing them afterwards. The calculations in the previous examples are based on simple geometry; however, there are many engineers who are not comfortable having to perform mathematical calculations, especially on site! For these engineers there is some good news. Various lens calculators, look-up tables and lens finders have been developed which for most applications do away with the need for any calculations. We shall look at these at the end of this chapter.

For many fixed cameras, an alternative to precision lens calculation is to use a *varifocal lens*. This offers a crude 2:1 adjustment range, typically, 4–8 mm or 6–12 mm. The principle of operation is very simple. Referring again to Figure 4.2, a varifocal lens has an adjustment which moves the centre lens assembly, making the focal length, and therefore the magnification, variable. However, it must be emphasized that this is not a true *zoom lens* because, every time the focal length is altered, the focus ring must also be reset. A zoom lens has a more complex optical assembly which compensates for the changing focal length and maintains correct focus throughout the zoom range. Zoom lenses will be discussed later in this chapter.

Many installers employ varifocal lenses because they allow for a margin of error and/or a change of mind on the part of the customer once the installation has been completed. On the other hand these lenses are more expensive than equivalent quality and size fixed lens and, if the lens is never going to be adjusted after the initial installation, a fixed lens is far more cost-effective and can make a quote for a system more competitive.

Another important consideration for a CCTV camera is the amount of light falling onto the pick-up sensor. Insufficient light results not only in a dark picture, but it will also be lacking in contrast. Excessive light will result in what is known as picture burnout. This is where lighter areas of the picture become overexposed, resulting in bright white patches with no detail whatsoever. In general, as long as the image is not allowed to burn out, the picture quality will be better where there is a high level of light input. The contrast ratio will be high, and the depth of field (we shall look at this later in the chapter) will be greatest.

Because a camera is expected to function effectively across a wide range of lighting levels, the lens must have some means of controlling the amount of light falling onto the camera pick-up. This is done by means of a mechanical *iris*. The iris is a delicate mechanism comprising a number of plates which slide as they are rotated. The principle is illustrated in Figure 4.8.

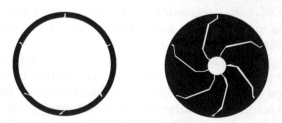

Figure 4.8 *Sliding plates are used to perform the function of an iris*

The *size of the hole created by the iris is known as the aperture*. As the diameter of the aperture changes, the change in the amount of light passing through follows a mathematical rule known as the inverse square law. If this sounds a bit of a mouthful, then consider it this way: *doubling the diameter of the aperture results in an increase in light throughput of 4 times (2^2); tripling* *the diameter of the aperture gives an increase of 9 times (3^3), and so on.* Similarly, halving the diameter of the aperture results in a light reduction factor of four, etc.

The light-gathering ability of a lens is known as its optical speed. The greater the amount of light falling onto the camera pick-up, the faster the CCDs will charge. Hence a lens with a high light throughput is said to have a fast speed. Larger aperture lenses will obviously be capable of gathering more light than smaller ones, and so in general they have a faster optical speed than smaller lenses. *The optical speed is stated as the F-number.*

The F-number is determined from:

$$F = \frac{\text{focal length}}{\text{lens/aperture diameter}}$$

If you look at the F-numbers on the side of a manual iris lens, you will see that they appear to have somewhat peculiar values. This is because *each position, known as an F-stop, is designed to give either a halving or doubling of light throughput*, and the reason why the numbers are at first glance unusual is down to simple geometry. To double the light throughput, the aperture *area* is doubled; however, to double the area of a circle, the diameter is not doubled; rather, it is increased by a factor of 1.4142. Likewise, to halve the area of a circle, the diameter is divided by a factor of 1.4142. Look at the following example.

For a 50 mm diameter lens with a focal length (FL) of 50 mm, the F-number would be 50 ÷ 50 = 1.

- In order to reduce the light-gathering area of this lens by 50%, the aperture diameter will be 50 ÷ 1.4142 = 36 mm.
- This gives an F-number of 50 ÷ 36 = 1.4.

Repeating these calculations for the 36 mm aperture area would give us a 25 mm aperture size, and an F-number of 2.0. So it can be seen that moving the aperture setting from the 1.4 position to the 2.0 position, i.e., one *F-stop*, reduces the light throughput by a factor of 50%. From this we can conclude that *each and every F-number represents a halving or doubling of light throughput*.

To illustrate this effect further, Table 4.2 shows the effect of each F-stop on the aperture diameter and light attenuation for a total of 13 stops on a 36 mm diameter, 50 mm focal length lens. F1.4 indicates that the aperture is fully open. In this example, it has been assumed that when the aperture is open, a light level of 2 lux is passing through the lens, and this level of light falling on the CCD chip in the camera is sufficient to produce an acceptable clear image. If the light input level increases, the aperture will have to close down accordingly to maintain a level of 2 lux at the CCD. Thus, for example, when the light input increases from 2 lux to 4 lux, the aperture diameter must reduce from 36 mm to 25 mm to maintain an output level of 2 lux at the CCD chip.

On a manual iris (MI) lens the aperture is adjusted by rotating a ring or collar which has the F-numbers indicated beside it, the letter 'C' marking the point where the aperture is completely closed.

From Table 4.2 notice that when the light level increases from 2 lux to 8000 lux, the aperture diameter changes from 36 mm to 0.6 mm, a total of

Table 4.2 *Effect of F-stops on aperture diameter and amount of light attenuation. It is assumed that the light output from the lens (at the CCD pick-up device) is 2 lux when the aperture setting is 1.4*

Aperture diameter	Area ÷ 2 (÷ 1.4142)	F-number	Light input (lux)
36 mm	–	1.4	2
36 mm	25 mm	2.0	4
25 mm	18 mm	2.8	8
18 mm	12.5 mm	4.0	16
12.5 mm	8.8 mm	5.6	32
8.8 mm	6.25 mm	8.0	64
6.25 mm	4.4 mm	11	128
4.4 mm	3.1 mm	16	256
3.1 mm	2.2 mm	22	512
2.2 mm	1.6 mm	32	1024
1.6 mm	1.1 mm	44	2 k
1.1 mm	0.8 mm	64	4 k
0.8 mm	0.6 mm	88	8 k

13 stops, in order to maintain a constant level of 2 lux at the pick-up. 8000 lux is what might be expected on a dark, overcast day. For this same lens to deliver 2 lux at the pick-up on a bright sunny day where the light level is typically 500 000 lux, the aperture would have to close down another 6 stops, and the diameter would become just 0.1 mm. Such a small diameter is simply not possible to control, either manually or automatically, because an error of just 0.05 mm would result in a light difference of 25% and, in the case of an automatic iris lens, the control circuit would 'hunt' (see 'Electrical connections' later in this chapter) causing the iris mechanism to oscillate. Clearly some other solution must be found.

The answer to the problem is to add a *neutral density* (ND) *spot filter* inside the lens. A neutral density filter is one which affects all frequencies of visible light by the same amount, and therefore has the effect of reducing the overall light level. The term 'spot filter' in this case refers to the fact that the filter does not cover the entire lens area but rather appears like a spot in the centre of the lens. The filter is designed to have maximum effect in its centre, reducing consistently between the centre and the outer edge. The principle is illustrated in Figure 4.9.

When the aperture is fully open the filter has little effect on the overall amount of light passing through. However, as the diameter of the aperture approaches that of the filter, its effect begins to be felt. Consider the effect that such a filter would have on the lens example we have been considering in

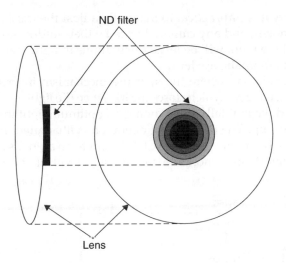

Figure 4.9 *ND spot filter on a lens*

Table 4.2. Let us suppose that the filter begins to take effect as the aperture diameter reaches 1 mm. As the light input increases from around 2000 lux to 4000 lux there is no need for the iris to close down as far to maintain a 2 lux output, because some of the incoming light is lost through the filter. Further increases in the incoming light level are met with corresponding amounts of filtering as the iris closes, and the centre area of the filter becomes the only part of the lens that is being used.

One option available to lens manufacturers for obtaining lower F-stop figures is to produce *aspherical lenses*. Without going into great detail, these lenses are ground in such a way that they are made to be more of a bell shape, which greatly improves the light-gathering and focusing ability. The increase in light efficiency enables lower F-numbers to be achieved, although this is at some considerable financial cost because these lenses are far more difficult to manufacture.

In practice, lenses using spot filters are often labelled 'F1.4–64'. This means that the mechanical iris offers 10 stops (F1.4–F32) unaided, with a further 11 stops (F32–F64) available in the filter ring area. Thus the lens assembly is capable of 21 stops without the aperture diameter having to close to an impractical amount.

The F-number can have serious implications when selecting a lens for a specific application. From Table 4.2 we see that a 50 mm lens with an aperture set at 25 mm equates to F2. However, for a zoom lens with a focal length of 150 mm, a 75 mm aperture would also give an F-number of $150/75 = 2$. So we see that every increase in focal length requires a larger aperture to maintain a workable lens speed.

In the case of a manual iris lens, the situation can arise where the picture quality is perfectly acceptable during the daytime, but as soon the light level begins to fall, the lens is unable to pass sufficient light to maintain image quality. Clearly it is impractical to manually adjust the iris every time the light level changes, and any camera located either outdoors, or indoors in a situation where there will be large fluctuations in light levels, should be fitted with an automatic iris lens.

An auto lens is one where the aperture mechanism incorporates some form of motorized drive which closes the iris when the light level is high, and opens it as and when it falls, maintaining an optimum light throughput for the camera pick-up. The principle of operation is illustrated in Figure 4.10. The video signal emerging from the CCD is sent to an auto iris amplifier circuit, which detects and monitors the average level of the video signal.

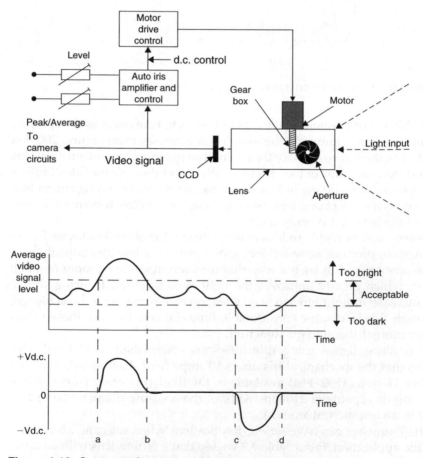

Figure 4.10 *Concept of motorized iris control*

Whilst the video level falls within a range deemed acceptable, nothing happens. However, should the light level increase, the average video level will rise, and the circuit will produce a d.c. output to activate the motor drive circuit, and hence the motor. The greater the light input, the higher will be the d.c. control, and the motor turns faster to close down the aperture.

This action is illustrated in the time-related graphs in Figure 4.10. At point (a) the light input becomes too high (large video signal), and the monitor circuit produces a positive d.c. output to operate the iris motor. Between (a) and (b) the iris is closing, the light level is falling, and the d.c. output from the monitor circuit falls accordingly until at (b), the aperture setting is correct and the motor stops. Between points (b) and (c) the light level, although varying, remains within acceptable levels, and the iris remains stationary. At (c) the light level suddenly falls, and the monitor circuit produces a negative d.c. which will cause the motor to turn in the opposite direction, opening the aperture. At (d) the light falling onto the CCD is once more within acceptable levels, and the monitor circuit removes the negative d.c. control, and the motor stops.

With the arrangement shown in Figure 4.10, the servomotor, gearbox, and auto iris amplifier circuit are all incorporated within the lens, which is commonly known as an AI lens. A connection is made from the camera to the lens to provide a video signal input to the monitor circuit. This link may use terminal connectors, or it may be made using a plug and socket arrangement. Some cameras have just one single socket with a selector switch labelled something like *'Video/DD Iris'*. We shall come to DD in a moment, but where the auto iris amplifier circuit is located in the lens assembly, the selector must be set to the 'Video' position.

Although AI lenses incorporating a motor and monitor circuit are used, they are very large and expensive. A less expensive alternative to the motorized drive is the *galvanometric drive*. The galvanometer is by no means a new invention; it was used as a movement in very sensitive moving coil measuring meters many years ago, and operates on very similar electromagnetic principles to the analogue meter movement which is still in use today. However, instead of moving a pointer needle, the galvanometer moves the plates of an iris mechanism. The principle is shown in Figure 4.11. The d.c. output is fed to the galvanometer plates, which in turn hold the aperture at the correct position; however, the d.c. must be maintained otherwise the plates will fall fully closed. Therefore the signal monitor circuit produces a varying d.c. as shown. In addition to the reduced cost, two advantages of employing galvanometric drive in preference to d.c. motors are a reduction in the physical size of the lens, and a much lower current consumption.

The lens size can be further reduced by incorporating the AI amplifier circuit within the camera. These compact lens types are known as *direct drive* (DD) lenses and are widely used in the CCTV industry. Where a DD lens is employed, the Video/DD selector switch on the camera must be set to the 'DD' position.

The two controls labelled 'Level' and 'Peak/Average' in the arrangement in Figure 4.11 give the installer some control over the iris action and thus the image on the screen. These controls may be found in either the lens or the camera, depending on where the AI amplifier circuit is located. They are required because there is no camera/lens combination yet devised which can automatically cope with every possible lighting scenario, and sometimes compromises need to be reached.

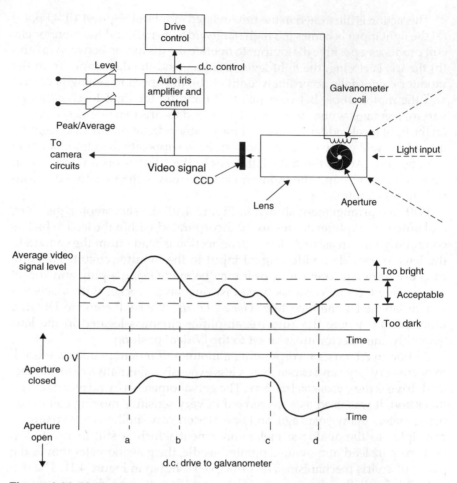

Figure 4.11 *Unlike the motor, a galvanometer requires a constant current to maintain the aperture setting*

The level control performs a similar function to the manual iris ring in that it allows the engineer to set the sensitivity of the lens. The control is usually labelled 'H–L' or 'Hi–Lo'. Setting the control towards the high end tends to produce satisfactory results in the day, but the picture may be dark at night. A low setting will allow the iris to open in the dark, producing a clearer picture. However, this may lead to excessive brightness in the day. Unless the system incorporates a facility whereby an operator can control the iris via telemetry, the level has to be set taking account of the extremes under which the lens will be expected to operate, and the particular requirements of the customer.

The peak/average control, often labelled 'P–A' or 'Pk–Av', enables the installer to introduce a degree of compromise with regard to large differences in lighting levels within the picture area. Consider the two situations

Figure 4.12a *Peak levels allowed to burn out; average levels are viewable*

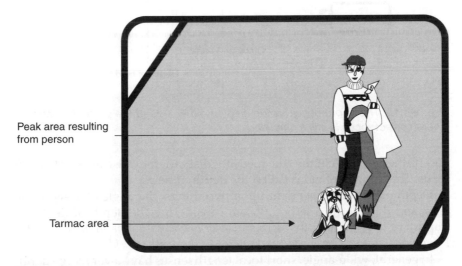

Figure 4.12b *Average area appears dark; peak areas are viewable*

shown in Figure 4.12. The wall light in Figure 4.12a would cause the iris to close down, resulting in a very dark image across the rest of the viewing area. In this situation the peak/average control may be used to open the iris up (by adjusting towards the 'average' position) in order to give a reasonable contrast across the picture. The compromise is that there will be a 'hot spot' in the area surrounding the wall light.

For the situation in Figure 4.12b, if the control were set for 'average', the iris would open fully to bring out the tarmac area, causing the person to bleach out. The problem would be particularly acute in low light conditions. Adjusting the control towards the 'Peak' setting causes the iris to close down to a point where the tarmac is not really visible; however, the person becomes more clearly visible on the screen.

You will frequently see another setting on many cameras labelled 'EI', meaning electronic iris. This is a circuit within the camera that maintains the correct video signal level using electronic means, and has nothing to do with the iris in the lens. This principle will be discussed in Chapter 6.

The final parameter we need to examine is the *depth of field. This is the range (in distance) in front of the lens where objects remain in sharp focus.*

As much as we would like a lens to display everything, both far and near, within the field of view in perfect focus, this is not possible. Look at Figure 4.13, where a lens is focused at a particular distance. Items closer

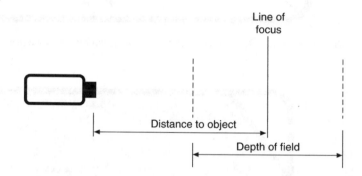

Figure 4.13 *The depth of field is the area on either side of the object in focus where correct focus is still maintained*

and further away than the true focusing distance may also appear to be in focus. This range of field is called the depth of field.

Depth of field is dependent upon two things: the angle of view of the lens, and the aperture setting. When considering the depth of field it is important to remember that either *a wider angle of view, or an increase in the F-number (reduced size of aperture), increases the depth of field.*

In general, wide-angle, short focal length lenses have a very large depth of field, even when the F-number is low. However, the advantage of the large depth of field is somewhat offset by the fact that distant objects appear very small on the monitor screen. As focal lengths increase, the lens views narrow and distant objects become larger. However, the depth of field reduces.

The problem now is that once the lens has been trained onto a target, and the focus has been set for that target, as soon as the lens is made to zoom onto a target at a different distance, the focus has to be re-adjusted. This is illustrated in Figure 4.14 where a zoom lens has been set to a target (object A) at distance D_1, and the focus has been adjusted accordingly. The depth of field around that object would be that shown by the dotted lines.

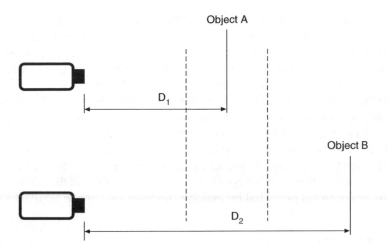

Figure 4.14 *When the zoom is altered, the focus will be incorrect*

However, when the focal length is re-adjusted to zoom in on object B at a distance D_2, the focus is incorrect. This is why the more powerful zoom lenses need to be re-focused when moved from one target to another.

Another related problem is when an auto iris, long focal length (tele photo) lens is used where there is insufficient lighting after dark. During the day the aperture will set to a high F-number, and there will be an acceptable depth of field. However, after dark the auto iris will reduce the F-number in an attempt to maintain the light throughput. When this happens the depth of field reduces. Of course, if the engineer only visits the site during the day he may wonder what the customer is complaining about!

Zoom lenses

Fixed focal length lenses are all right for locations where the area we wish to view is only a few metres square. However, as soon as we attempt to view a larger expanse such as a car park or shopping mall, then the limitations of a fixed lens become patently obvious. To attain the field of view a wide-angle lens must be used, but then everything and everyone appears so small on the screen that the image is useless for evidential purposes. On the other hand, if a zoom lens is fitted the field of view is lost meaning that, unless the lens is motorized and can be operated from a control room, it is of little use. This in turn means having to employ staff to operate the CCTV system. For situations where the level of risk and the size of the business where the system is to be installed does not warrant employing full-time CCTV operators, let alone invest in expensive zoom lenses and telemetry systems, it is possible to obtain a reasonable coverage by installing two cameras at the same location, one fitted with a wide-angle lens and the other with a zoom lens. However, the installer must be certain that such a system will provide the level of cover required before selling the idea to the client.

It is common practice to state the performance of a zoom lens in terms of the ratio of the change in focal length. For example, a lens having a focal length from 16 mm to 160 mm would be quoted as being a 10:1 zoom lens. However, just describing the ratio gives no clue as to the range of views provided; i.e., 10:1 could mean 10 mm to 100 mm or 15 mm to 150 mm, therefore manufacturers usually quote the actual focal length range as well as the ratio.

The principle of a zoom lens is shown in Figure 4.15. Within the assembly, the zoom lens group is the most complex part of the mechanism because the optical devices within this group have to move in such a way that the image remains undistorted and in correct focus. It would be very difficult to maintain correct focusing over a large change in focal length, and therefore a manual focus control is provided by making the front lens group adjustable.

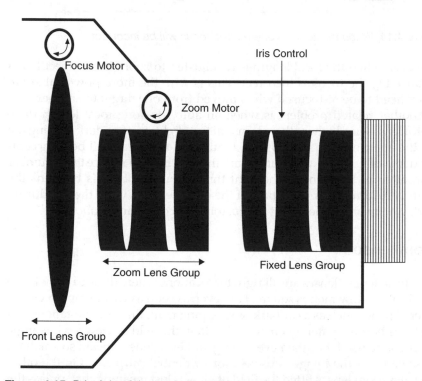

Figure 4.15 *Principle of the zoom lens. In a manually operated lens, the two motors are omitted and two small levers are available to turn the focus and zoom controls*

Specifying the optical speed of a zoom lens is not as straightforward as for a fixed lens because, as shown earlier, the F-number is dependent on the focal length and aperture, and in this case the focal length is variable. To keep the specifications simple, the majority of manufacturers simply state the F-number for a lens at the two extremes of its focal length. So, for example, a ⅔" 6:1 zoom lens having a focal length range between 12.5 mm and 75 mm may be quoted as having an F-number between F1.2 and F560.

When a zoom lens is set to its maximum focal length (telephoto), it must be fixed firmly in position because the slightest movement results in a very large picture shake on the monitor. The longer the focal length, the more acute the problem. Where the camera is fixed to a solid structure there is not usually a problem; however, when it is mounted on top of a tower, the slightest wind can throw the picture all over the monitor screen when the lens is at a high telephoto setting. Gyroscopically corrected lenses are available (at considerable cost). The optics are mounted in a mechanism that moves in the opposite direction to the motion of the main lens casing. Needless to say, because of their high cost these devices are not commonplace.

Electrical connections

Correct electrical connection between the various lens drive circuits and the lens is very important. Connection methods, terminology, and drive potentials can differ between lens types and manufacturers, and an engineer must be able to decipher the wiring diagrams provided in the manufacturers' technical information.

The most popular lenses for use in smaller installations are either fixed focal length or varifocal types, employing a DD or an AI iris control. In this case the only electrical connections between camera and lens are for the iris. Terminal connectors may be used on some cameras; however, many lenses come with a prefitted connector plug. When using one of these with a camera from the same manufacturer, it is usually very difficult to get the connections wrong. However, where a mix of equipment is being installed it is important to check for compatibility because, although some standardization has taken place with the acceptance of the 'P Plug' (illustrated in Figure 4.16), there are still a number of wiring configurations. It is also

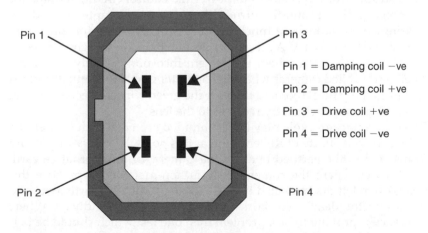

Figure 4.16 *Standard connection configuration for the P-Plug. (So called because it was developed by Panasonic)*

essential that, if fitted, the Video/DD selector switch on the camera for iris control output is correctly set.

A galvanometric iris usually has a four-pin connector. Two terminals connect the d.c. drive voltage from the camera to the actuator coil which controls the aperture. The other two carry a control voltage to a second coil in the iris assembly, which is used for damping. The damping coil is necessary because the galvanometer is a very sensitive device and is prone to over-reaction to the drive voltage. Because the iris control is effectively a closed loop servo (the light input forming the feedback), this would result in a condition known as hunting. Hunting is where a servo continually overcompensates and the unit that it is controlling, in this case the iris, oscillates. Thus the iris control circuit in the camera outputs not only the drive voltage but also a damping voltage which is fed to the damping coil. The arrangement is shown in Figure 4.17.

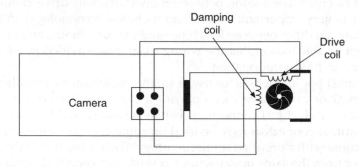

Figure 4.17 *Connections between camera and a DD auto iris lens*

In cases where the iris control circuit is located within the lens, three connections are required: one for the video signal, one to provide a 12 Vd.c. supply from the camera to power the control circuit, and one for a common negative (ground) return. Although not a guaranteed standard, the wiring colours for these functions are normally white for video, red for +12 V and black for 0 V. A green wire is often found in the lens connector cable. This is only used where the system incorporates fully functional telemetry-controlled cameras where the operator is able to manually adjust the iris from the control room. In this case the green wire connects a varying d.c. control from the telemetry receiver to the lens.

The lens connector cable may have a braided screen, which is intended to reduce the possibility of RFI entering at this point. In theory this would be connected to the earthed body of the camera, or an alternative earth point in cases where the camera body is non-metallic. In practice this screen is often left disconnected because it can produce an earth loop, and as the connecting lead is so short it is rare for RFI to be induced. Where interference is proving to be a problem then one test which should be performed would be to connect the screen to earth.

When we come to look at motorized zoom lenses, the electrical connections become a little more involved. A zoom lens usually employs low-voltage d.c. motors to move the zoom lens group (focal length) and the front lens group (focus). Typical operating voltages for these motors can be between 5 and 12 Vd.c., the control voltage being provided by the telemetry controller, which may vary depending on the type of system employed. It is important to check that the lenses selected for use with a control system are compatible. For example, if the lens employs 5 V motors, and the control unit is applying 12 V when zoom and focus is operating, you will find that the lenses might react very quickly(!), but also fail soon after installation because the motors have burnt out. On the other hand, operating 12 V motors at 5 V will mean that the lens reaction is very slow, if there is any reaction at all.

The motors drive the lenses via gearboxes and must not be allowed to over-drive the mechanism, otherwise permanent damage may occur. Take for example the zoom mechanism. When the motor has moved the zoom lens group to the maximum focal length position, the power to the motor must be cut immediately. This can be achieved using *limit switches*. However, there is no point in simply placing a switch in series with the motor, other-wise once it has opened there is no way of applying a reverse voltage to move the lens in the other direction. The solution is to place diodes across the switches. A typical circuit arrangement in shown in Figure 4.18.

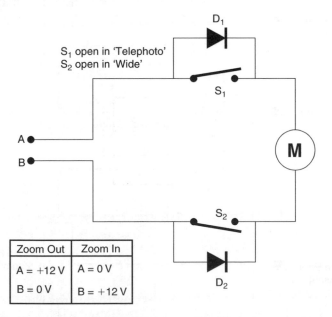

Figure 4.18 *Limit switch circuit for a zoom motor. The same arrangement can be used for focus and iris motors*

In this circuit, switch S_1 opens when the lens reaches maximum zoom (telephoto). When the operator wishes to zoom out again it is necessary to reverse the motor direction, which means reversing the polarity of the drive voltage at terminals A and B. The current path will now be from terminal A, through diode D_1, through the motor, returning through switch S_2, which will still be closed. Moving from the 'wide' position is simply the reverse, because S_2 will now be open, and diode D_2 provides the current path.

It is not uncommon to find potentiometers in a zoom lens for *preset zoom* and *preset focus*. These potentiometers rotate as the zoom and focus motors are operated, thus deriving a range of voltages corresponding to the zoom/focus ranges. The output voltage from each of the potentio-meter sliders is connected back to the control unit, making possible a facility whereby the control unit can have programmable preset zoom positions. A typical potentiometer wiring arrangement is shown in Figure 4.19.

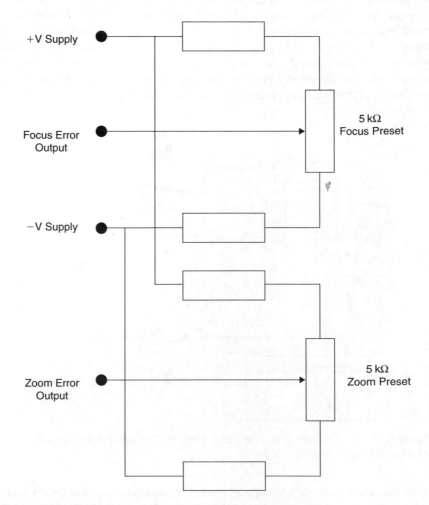

Figure 4.19 *Preset control circuit arrangement*

When setting up the preset positions, the engineer or operator moves the zoom to a desired position (it may be that the pan/tilt is also moved at this point) and adjusts the focus. This position can then be stored into the memory of the control unit, which does this by converting the d.c. value from each potentiometer into a digital value. During normal operation, when the operator instructs the control unit to move the camera to the preset position, the zoom and focus motors (plus pan/tilt motors) are operated until the d.c. from the potentiometers reaches the predetermined value.

Modern control units can memorize more than one preset position; indeed, some can store up to one hundred presets. Yet in most cases five is about the most that is required because, when you think of it, when would you need one camera to have a hundred zoom positions, and how can an operator remember where they are anyway?

Lens mounts

A type of lens mount referred to as *C mount* has been used for many years in the cine industry, and because these lenses were already in manufacture they were adopted by the early CCTV industry. The term 'C mount' refers to the type of screw thread on the lens, and the distance from the back of the lens to the CCD pick-up device in the camera.

C mount lenses are comparatively large, and as the size of CCTV cameras continued to reduce, it was necessary to have a more compact lens. Furthermore, the lenses can be one of the most expensive components in a CCTV system, and as the size of systems increased in terms of the number of cameras, a more cost-effective lens was called for. It was with these factors in mind that the *CS mount* lens was developed.

The CS mount lens is much smaller; however, this is not without some sacrifice in performance. Compared to C mount lenses, CS mount types suffer greater optical distortion because nearly all of the glass area is employed, whereas a C mount only makes use of the inner area where there is minimal distortion.

From an engineer's point of view, the main difference between these lenses is the distance from the back of the lens to the image device in the camera. *For a C mount this distance is 17.5 mm, whereas for a CS mount it is only 12.5 mm.* In practice this means that these lens types cannot simply be interchanged because the focal point will not fall onto the pick-up device.

CCTV cameras are specified as being either C or CS mount, although some have a mechanical adjustment which moves the pick-up device between 12.5 mm and 17.5 mm. Where the mount cannot be adjusted, the following rules apply.

A C mount lens can work on a CS mount camera as long as a special 5 mm adapter ring is fitted to extend the distance to 17.5 mm. This is

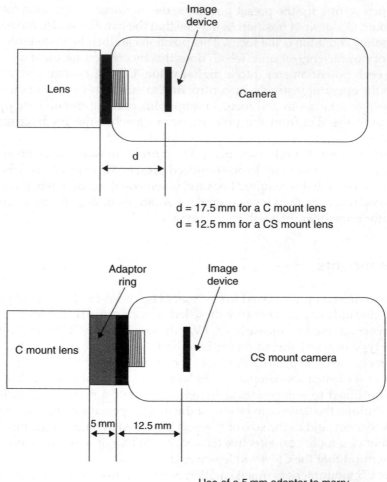

d = 17.5 mm for a C mount lens
d = 12.5 mm for a CS mount lens

Use of a 5 mm adaptor to marry
a C mount lens to a CS mount camera

Figure 4.20

illustrated in Figure 4.20. However, a *CS mount lens will never work on a C mount camera* because there is no way of reducing the 17.5 mm distance.

Filters

There are occasions where it is useful to fit a light filter to the lens. For example, where a camera is required to look through a glass windowpane, a *polarizing filter* can be fitted to remove the effects of glare or reflections. This type of filter is fitted to the lens and then rotated until the desired result is achieved.

Another type of external filter is the ND type, which we discussed earlier in this chapter when considering the action of the spot type filter, which is used to increase the possible number of F-stops. Full-sized versions affecting the entire lens area are available, and in a range of attenuation factors. However, it is not often in CCTV applications that these would be a permanent fixture to a camera. One practical application of the ND filter is in the adjustment of back-focus, which will be discussed later in this chapter.

Infrared cut filters are used where it is desirable to remove IR light from the camera. All colour cameras have an in-built IR cut filter; however, some monochrome models are designed to respond to IR light. Sometimes where there is a lot of infrared light in the viewing area it is necessary to negate this feature because parts of the picture become slightly burnt out. Fitting an external IR cut filter often results in a picture with an improved grey scale contrast range. Another application for IR cut filters is when adjusting the focus. Removal of infrared light from the lens enables the focus to be adjusted using only visible light. This produces a slightly different focusing point, but one which is perhaps more accurate for a camera which is expected to rely largely on visible light.

Opposite to the IR cut filter is the IR pass filter, which removes the majority of visible light so that the camera can only 'see' infrared. In the case of cameras that are intended primarily for night-time operation using infrared lighting, adjustment can be performed in the daytime if an IR pass filter is used to remove the visible light. The reliability of the adjustment can be further improved by adding an ND filter, in addition to the IR pass filter, to further block visible light.

Lens adjustment

One of the biggest problems in CCTV installations is incorrect adjustment of the *back-focus*. This refers to the 17.5 mm/12.5 mm distance between the back of the lens and the camera pick-up device. Where fixed lenses are involved this is not as critical as long as the engineer is able to obtain a correctly focused image. However, when a zoom lens is employed, if the back-focus is not correctly adjusted, the operator will find that the focus moves out every time the zoom function is used. Another problem can be that the zoom lens functions satisfactorily during the day when the auto iris is closed down, but when the light begins to fade and the iris opens, the focus becomes poor due to the reduced depth of field.

We already touched on this problem earlier when we discussed depth of field, and the ideal solution is to perform back-focus adjustment at dusk when the adjustment is far more sensitive. However, in reality this is not always practical, and therefore another method is to fit an external ND filter to the lens whilst performing the adjustment – the filter simulating twilight conditions by reducing the amount of light input.

There is more than one acceptable method for obtaining correct back-focus adjustment, and engineers will adopt their own preferred routines as they become proficient at this. However, when starting out, the generally accepted method for adjusting the back-focus on a camera which is fitted with a zoom lens is as follows.

1. Manually open the iris, or fit an ND filter, or work in low light conditions.
2. Select a target at the maximum operational range for the particular camera/lens (a target with a lot of detail such as a wall is ideal).
3. Adjust the lens focus to 'far'.
4. Set the zoom to maximum wide angle.
5. Move the back-focus adjustment on the camera forwards and back until optimum focus is obtained.
6. Set the zoom to full telephoto.
7. Adjust the lens focus for optimum focus.
8. Set the zoom to maximum wide angle.
9. Re-adjust the back-focus for optimum focus.
10. Repeat steps 4 to 10 until optimum focus is obtained at all points between wide and telephoto.

In some cases the back-focus is fixed with a locking screw or nut. Be sure to slacken this off before commencing adjustment, and be sure to secure it again afterwards, otherwise vibrations from the zoom or PTZ action may move the back-focus adjustment.

From a practical point of view, it has been assumed all along that the engineer is looking at the picture on a monitor whilst performing focus adjustments. However, this is not always as straightforward as it might seem, especially when working on a camera assembly mounted atop a 7 m tower. There are a number of ways of overcoming this issue. One of them is to adjust the back-focus before mounting the camera, perhaps even in the workshop. When performing the adjustment in this manner, remember to reduce the lighting somehow, otherwise you might find yourself performing the adjustment again once the camera has been installed.

Another approach is to use a portable, battery-powered monitor. There are test monitors made specifically for this purpose, and they can prove very useful in the field for performing all manner of tests and adjustments. Difficulties are sometimes experienced when using these in direct sunlight, as viewing can be somewhat difficult.

Another device that is useful for performing focus adjustment is the focus meter (Figure 4.21). This is designed to connect to the video output from the camera, where it analyses the high-frequency video components in the signal. The sharper the focus the greater the amount of h.f., and thus if the engineer moves the focus adjustment to one end of its travel and then progressively moves it back once again, the meter will indicate an increasing h.f. content up until the point where the optimum focus point has been passed, whereupon the indication will begin to fall off. The engineer is therefore able to set the focus adjustment for the peak indication.

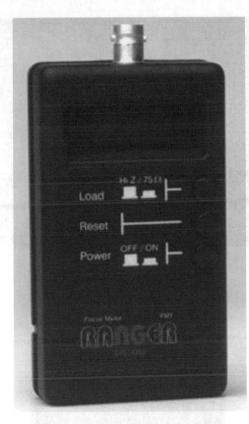

Figure 4.21 *A focus meter. (Courtesy of NG Systems Ltd)*

Lens finding

Earlier in this chapter we saw how we can calculate the required lens size for a given application; however, there are alternatives to using mathematics and an electronic calculator. Perhaps the simplest of these is to use one of the lens calculators which are often available free of charge from lens manufacturers or suppliers. These differ slightly in both the way that they function, and the data they produce, but they are simple to use, are quite accurate and are adequate for the majority of applications. A typical lens calculator is illustrated in Figure 4.22. This particular type requires you to know the primary target height or width, and the distance from the camera to the target. By adjusting the cursors on the calculator, the focal length and the horizontal or vertical angle of view can be found for any given camera format size.

Another method is to use one of the look-up tables produced by various lens manufacturers; however, these often quote the horizontal angle of view produced by a range of lenses when fitted to different camera

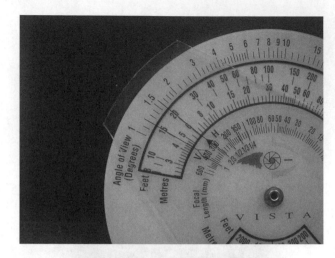

Figure 4.22 *Typical lens calculator*

formats. In this case the engineer still has no idea of the object width or height for any given distance from the lens, but this can be calculated using trigonometry:

$$W = d \tan \theta$$

where W is the width of area across the image, d is the distance of object from lens, and θ is the horizontal angle of the lens.

Example

A look-up table states that a 30 mm, ½″ format lens has a horizontal angle of approximately 11°. Find the image width at a distance of 40 m.

Solution

$$W = 40 \times \tan 11°$$
$$= 40 \times 0.19$$
$$= 7.6\,\text{m}$$

Of all the tools available for lens finding, perhaps the most effective are the optical viewfinder types, an example of which is shown in Figure 4.23. The engineer positions himself where the camera is to be located and

Figure 4.23 *Optical device used for determining the required lens focal length for a given field of view, target distance and camera format size. (Photo courtesy of CBC (Europe) Ltd, manufacturers of Computar lenses)*

looks through the device, adjusting it until the desired image is obtained. Calibrated markings on the side of the device enable the focal length to be read off. The scales include the various format sizes. Apart from removing the need for any calculations, another bonus of using an optical viewfinder is that the customer can be shown what he is getting before agreeing to the installation. Indeed, the customer can set the viewfinder to show the engineer/surveyor what he requires, and if the engineer is seen to log the agreed lens size at the time, he need have no worries about comeback from the customer at a later stage in the installation.

5 Fundamentals of television

Before we can look at the operation of cameras and monitors, it is important to have an understanding of the makeup of the television picture, and the signals required to produce the picture. Later, in Chapter 7, we will examine the various picture display device technologies.

Development of television really began during the 1930s with a number of ideas being tested, with differing degrees of success. However, all of the ideas had one thing in common: they were designed to produce a picture using a *cathode ray tube* (CRT), and consequently modern analogue television systems are still tied to a video signal waveform that is designed to drive a CRT, even when a flat panel display device of some type is being used. Of course, digital signal transmission systems are replacing analogue in both the broadcast and CCTV arenas but, nevertheless, analogue signals will still be employed for a number of years to come because of the large amount of analogue equipment still in use and the prohibitive cost of replacing all of this overnight (not to mention the fact that manufacturers could not produce the amount of equipment necessary for a sudden upgrade in such a short period of time).

Producing a raster

A raster is the term given to the blank white display produced when a CRT is made to scan its screen without any video signal input. It is constructed by making the spot produced by the electron beam in a monochrome CRT (or three spots in the case of a colour CRT) deflect vertically and horizontally at high speed across the screen. The spot is moving so fast that it is indiscernible to the eye, and hence the brain is tricked into thinking that it is seeing a solid white display. The principle of sequentially scanning the screen to produce a raster is illustrated in Figure 5.1. For simplicity a raster with only nine lines has been drawn, although, as we shall see later, both broadcast TV and CCTV in the UK use a 625-line raster (525 lines for NTSC systems).

The horizontal (line scan) speed is made to be many times faster than the vertical (field scan) speed, and so the spot zigzags its way down the CRT screen. However, the *line scan* period from left to right is much longer than the *line flyback* period, where the spot moves rapidly from right to left. Similarly, the *field scan* period is many times longer than the *field flyback* period, where the spot is made to return very quickly to the top of the screen.

During both the line and field flyback periods the electron beam is cut off, and so only the solid lines numbered 1 to 9 in Figure 5.1 are actually

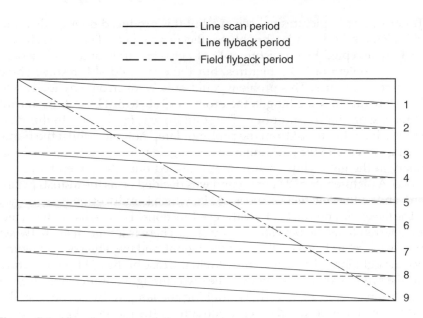

Line scan period
Line flyback period
Field flyback period

Figure 5.1 *Nine-line sequentially scanned raster*

displayed. This is why the flyback periods have been shown dotted in the illustration. The line scan period is often referred to as the *active line period*.

The electron beam in a CRT is deflected by passing currents through scanning coils that are wrapped around the CRT neck. During the flyback periods the currents are reversed to produce an equal but opposite magnetic field. This cannot be done instantly because of the inductive effects of the scan coils – hence the reason for having to turn off the beam and wait for flyback to complete.

So we see how a nine-line raster is produced using sequential scanning. But what scanning speeds would we require to produce a practical 625-line or, for NTSC, a 525-line raster? The answer to this question is not as straightforward as it may appear. During the development of the television system there were a number of factors the designers had to take into consideration when determining the scanning speeds. Primarily these were the picture rate, the picture resolution, the system bandwidth, and the problem of picture flicker.

It must be remembered that the television system had to be designed around the existing cinema system, because at the time it was accepted that much of the transmitted material would be taken from movie sources. Even television newsreel was sourced on celluloid film (there were no camcorders in the 1940s!). So, before we can begin to look at the rationale for constructing television pictures using a CRT, we need to look briefly at the structure of the source material that was to come from the cinema industry.

The 'moving pictures' we see at the cinema are actually a series of still photographs being flashed onto a screen at high speed. Early systems used a frame rate of 16 pictures per second, but this produced poor quality and jerky pictures, and a 16 Hz brightness flicker was very discernible. It was generally accepted that a picture rate of between 20 Hz and 25 Hz would cure the problem of jerky pictures, but the eye could still resolve some flicker at these rates. This problem could only be overcome by increasing the picture rate to something in the order of 45 Hz to 50 Hz, but this meant a doubling of film consumption. The solution was (and still is to this day) to block the light output from the projector lamp not only whilst the frame is being pulled into the gate, but also once again whilst it is being held stationary in the gate. Thus each frame is flashed onto the screen twice per second. A picture rate of 24 per second was decided on as the industry standard, and so the cine flicker rate is 48 Hz, which is indiscernible to the eye.

The flicker is a product of how the eye functions. The eye has a characteristic referred to as the persistence of vision. There is a time delay between a part of the retina being excited by a light input, and the retina output fading once the light input has been removed. In the case of the 24 Hz cine picture rate, the eye was able to distinguish the projector light output being strobed as the frames were advanced. By strobing the lamp at twice the rate, the retina is unable to respond quickly enough for the flicker to be resolved.

Because the UK mains frequency is 50 Hz, there were reasons relating to receiver design why a picture rate of 25 per second rather than 24 was chosen for the television system. (In the USA, where the mains frequency is 60 Hz, the picture rate is 30 per second.) However, displaying the pictures at a rate of 25/30 Hz using sequential scanning would produce the same flicker problems as a cine film flashed at 24 Hz. To overcome the problem of picture flicker in television, a technique known as *interlaced scanning* was developed. The principle of interlaced scanning is illustrated in Figure 5.2 where, again for simplicity, a nine-line raster has been shown.

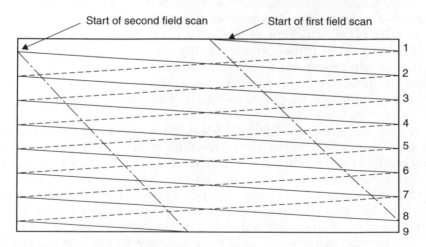

Figure 5.2 *Nine-line interlaced raster*

In this case the spot begins at the top centre of the screen and makes its way down, finishing at the bottom right-hand corner. The field flyback now returns the spot to the top left-hand corner, where it once again scans downwards to end in the bottom centre. During the first vertical scan period, known as the *odd field*, the odd TV lines are scanned, and 50% of a picture frame is displayed. During the *even field* period when the even line numbers are scanned, the other 50% of the frame is reproduced. So we see that *two TV fields equals one TV frame*. The principle is illustrated in Figure 5.3 where for simplicity an eleven-line raster has been used.

Odd field (312.5 lines PAL)   Even field (312.5 lines PAL)   1 TV frame (625 lines PAL)
  (262.5 lines NTSC)   (262.5 lines NTSC)   (525 lines NTSC)

Figure 5.3 *Two interlaced fields build up a complete TV picture frame*

The flicker is eliminated because all areas of the screen are being illuminated at a rate of 50 Hz (60 Hz NTSC). At a normal viewing distance the eye cannot discern the individual TV lines, and so just as one line would be appearing to fade, two more are scanned, one on either side.

In the PAL UK TV system (see later in this chapter), the vertical scanning rate is made to be 50 Hz, twice the 25 Hz picture rate. Thus during the first field scan the spot reproduces the odd 312.5 lines, and during the second field scan the even 312.5 lines are reproduced. So in 1/25th of a second (i.e., 40 ms or 25 Hz), a complete picture is built up on the screen.

For the NTSC system, the vertical scanning rate is made to be 60 Hz, twice the 30 Hz picture rate. Thus during the first field scan the spot reproduces the odd 262.5 lines, and during the second field scan the even 262.5 lines are reproduced. So in 1/30th of a second (i.e., 33 ms or 30 Hz) a complete picture is built up on the screen.

Returning to the question, 'what scanning speeds would we require to produce a practical 625-line raster?', having dealt with the issues of picture rate and picture flicker, let's now see where picture resolution and system bandwidth come in.

Picture resolution

This relates to the definition of a TV picture, and can be expressed in terms of either vertical or horizontal resolution. The vertical resolution is determined largely by the number of lines that make up the picture. The horizontal resolution is determined by the number of black and white elements that the system is capable of displaying along any line. Ideally, the maximum horizontal resolution for a television system should be at least equal to the vertical resolution – that is, where the width of each black or white square is equal to the thickness of one TV line. This display situation is illustrated in Figure 5.4.

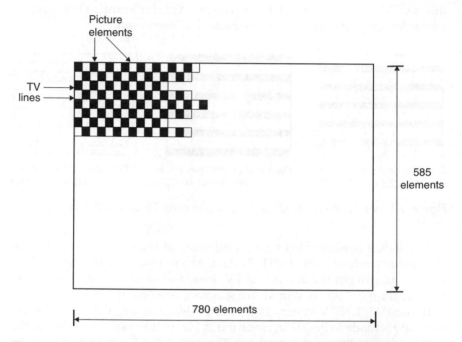

Figure 5.4 *Deriving horizontal resolution for a 625-line picture. (Pixels are not drawn to scale as they would appear too small)*

Although the vertical resolution is determined by the number of TV lines in the picture, it is also dependent on the position of the camera in relation to the image in view. The more horizontal lines a system has, the more squares we can reproduce. However, it does not necessarily hold that a 600-line raster would always give a 600-line resolution. Looking at Figure 5.4, consider what happens when the camera is positioned such that each line is scanning equally between two sets of squares. The line cannot be black and white at the same time, and so the entire checkerboard would appear mid-grey. This is an extreme situation, but it does make the point that in any picture there will almost certainly be some loss in vertical resolution.

Ignoring widescreen television, which has yet to make an inroad into CCTV, all monitor screens have an *aspect ratio* of 4:3. That is, although the screen size of any monitor, e.g. 30 cm, 51 cm, is measured diagonally, the ratio of the sides is always 4 wide by 3 high. Because the horizontal resolution must be at least equal to the vertical resolution, for any given number of

TV lines (vertical resolution) the horizontal resolution will be the number of TV lines multiplied by 4/3. Let's see how this works for the 625-line system employed in UK broadcast and CCTV.

As we have seen, a 625-line raster is constructed from two fields, each having 312.5 lines. The period for one field scan is 20 ms; however, this includes the flyback period which is (approximately) 1 ms, during which time the electron beam is cut off to prevent it being seen moving up the screen. In 1 ms approximately 15 horizontal lines will have been scanned, but the beam is switched off for 20 lines to ensure a clean flyback. In other words, out of the 312.5 lines per field, only 292.5 are active.

This means that over two field periods 40 lines are unused and so, out of 625 lines, only 585 actually contain picture information. Looking again at Figure 5.4 we see that if the vertical resolution is 585 lines, then the horizontal resolution should be at least 585 × 4/3 = 780 pixels.

We can now turn things around and look at them from a different point of view and ask the question: why was 625 chosen for the number of lines? It was derived from the decision to begin the calculations with horizontal resolution 780 pixels. So how was the figure of 780 arrived at? This is the maximum frequency of the signal drive voltage required at a CRT to reproduce the checkerboard display, a portion of which is shown in Figure 5.5.

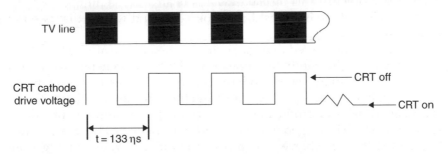

Figure 5.5

In the UK TV system, the horizontal scanning speed is set at 64 μs; 52 μs for the scan period (also referred to as the active line period); and 12 μs for the flyback period. From the periodic time of 64 μs, we can derive the scanning frequency:

$$\text{frequency} = \frac{1}{\text{periodic time}} = \frac{1}{64\,\mu s} = 15\,625\,\text{Hz}$$

For a display device to switch on and off quickly enough to reproduce 780 pixels during the 52 μs active line period, the periodic time for one cycle of the signal would be 133 ns. In terms of frequency this equates to 7.5 MHz. In other words, in order to reproduce a horizontal resolution of 780 pixels, the system including everything from the camera to the display device would require a bandwidth of 0 Hz to 7.5 MHz. This exceeds the

5.5 MHz bandwidth originally allocated to broadcast TV in the UK, but it was found that the bandwidth could be reduced by applying something called the Kell factor. This is a figure of 0.7, which was derived as a result of extensive work in 1933 by Ray Kell who, after performing many viewing tests, concluded that a reduction in horizontal resolution of 0.7 would not produce an appreciable deterioration in picture quality.

Applying the Kell factor reduces the bandwidth to a more practical figure of 5.5 MHz, and so it is that the UK TV system has a video bandwidth of 0–5.5 MHz, which equates to a resolution in the order of 546 pixels.

For NTSC the horizontal line frequency is 15 734 Hz, which produces a line period of 63.6 µs. The horizontal flyback period is approximately 10.3 µs, which gives an active line period of around 53.3 µs. The vertical flyback period is typically 15 lines per field (30 lines per frame), resulting in 495 active lines per TV frame. Therefore, applying the rule that the horizontal resolution should be at least equal to the vertical resolution, an ideal system should be capable of reproducing $495 \times 4/3 \times 0.7$ (Kell factor) = 462 pixels along one line. This would equate to a video signal bandwidth of 0–4.3 MHz. In practice the video bandwidth for NTSC is specified as being 0–4.2 MHz, which would produce a horizontal resolution of around 450 pixels.

It is more common to express horizontal resolution in terms of 'television lines' or *TVL*. This measurement is related to the number of horizontal pixels that we have just been looking at, and will be discussed in Chapter 6.

It must be remembered that the figures quoted above are for broadcast television and, whilst in times gone by, CCTV has in general fallen short of these specifications, with modern technology cameras and monitors, coupled with fibre-optic or other broadband transmission techniques, CCTV is actually capable of producing a superior image resolution to broadcast television, which is dogged with bandwidth restrictions. However, it must be stressed that to achieve a high resolution in CCTV, everything from the lens to the monitor must be of a high specification and the installation must be sound; one weak link in the system, for example poor cabling with a multiple of connectors, or a poor-quality lens, will result in an overall degradation of resolution.

Synchronization

The image which has been focused onto the pick-up device by the lens is scanned at a rate of 50 Hz, and 15 625 Hz (PAL). It is essential that the monitor scans at the same rate as the camera, not just in terms of frequency, but also in terms of the precise position of the spot on the screen. (Remember that, even though modern solid-state devices such as camera CCD chips and flat panel monitors do not actually scan, the analogue video signal must still conform to the CCIR standards that relate to the old camera tubes and CRT monitors. Hence the reason why we still talk in terms of scanning.)

Consider the two conditions illustrated in Figure 5.6. In condition A, the electron beam in the monitor begins to scan a field at precisely the same time that the camera begins a field scan, so the displayed picture appears on the screen in exactly the same position as it would if you were looking directly through the camera lens.

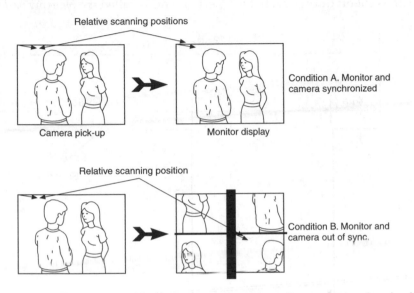

Relative scanning positions

Condition A. Monitor and camera synchronized

Camera pick-up

Monitor display

Relative scanning position

Condition B. Monitor and camera out of sync.

Figure 5.6 *Effect on the display when camera and monitor are unsynchronized*

In condition B, the beam in the monitor CRT is in the centre of the screen when the camera begins a field scan. In this case the top left corner of the image is displayed in the centre of the screen, with a corresponding displacement of the rest of the image. The vertical and horizontal dark stripes appear as a result of the camera flyback periods being displayed; remember that during the flyback period the beam is cut off, and so the camera outputs a black level signal.

The monitor must be synchronized to the camera, and to achieve this the camera generates a series of pulses which are added to the video signal at the output. At the end of each horizontal scan, at the instant that the camera is initiating line flyback, a *line synchronization* (sync) *pulse* is generated by the camera and added to the video signal. Likewise, at the end of each field scan a *field sync pulse* is added to the video signal. Thus, during one 20 ms field period the camera will output 312 line sync pulses and 1 field sync pulse. During one complete TV frame (40 ms) a total of 625-line, and two field sync pulses will be output from the camera.

The shape and timing of the line and field sync pulses is complex, and it is beyond the scope of this book to look into the reasons behind this complex make-up. However, it is important that a CCTV engineer can recognize these signals when they are viewed on an oscilloscope.

The line sync pulse is shown in Figure 5.7. The monitor will initiate line flyback at the instant the first falling edge, immediately following the front porch, of the pulse appears. Note that the total duration of the pulse is 12 μs, which is equal to the line flyback period. The porches are set to be at a level in the video signal waveform equal to black to ensure that the beam is cut off during the flyback period. This is called the *blanking period*.

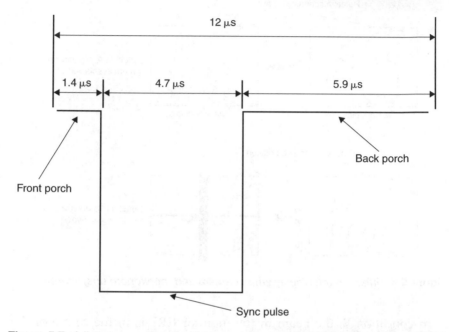

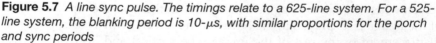

Figure 5.7 *A line sync pulse. The timings relate to a 625-line system. For a 525-line system, the blanking period is 10-μs, with similar proportions for the porch and sync periods*

When viewed on an oscilloscope at the output from a camera, the field sync pulse actually appears nothing like a pulse at all. This is because it is made up from a series of pulses beginning with five equalizing pulses, followed by five pulses which we refer to as the field sync period, and finally five or four (depending whether it is the odd or even field) more equalizing pulses. At the monitor, after separation from the video signal, this series of pulses is passed through a low-pass filter which integrates them into a single pulse, which is the field sync pulse. The complete field sync period is shown in Figure 5.8.

Following the field sync period, a series of black lines is sent out. These are essential to ensure that the beam is cut off during the field flyback period. As discussed previously, there are *twenty lines in the field-blanking period (15 for NTSC)*. In a UK broadcast transmission, it is during this period that Teletext information is transmitted and, although this might not appear significant for CCTV engineers, the idea of transmitting data during the

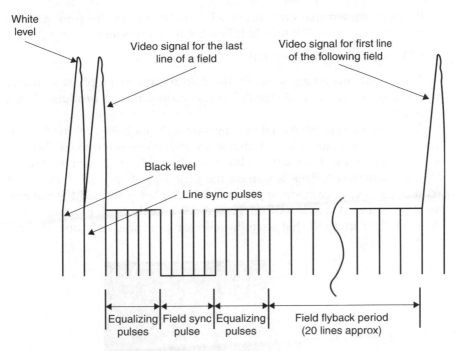

Figure 5.8 *Field synchronizing and flyback period*

field flyback period has been taken up by the majority of manufacturers of CCTV telemetry control equipment, and camera control data is sent out during this period in a very similar way to broadcast Teletext data. This will be considered further in Chapter 10.

Synchronization is very important and any loss, distortion or attenuation of the sync pulses will result in vertical jittering or rolling, and/or horizontal pulling or rolling. Such problems are all too common in CCTV installations, and can be the result of many things. To mention but a few, there can be different earth potentials between various points in the system, faulty or incorrectly installed cables and/or connectors, incorrect cables or connectors, poor-quality camera switching units which interrupt the sync signals, and incorrect positioning of terminator switches on monitors or other equipment. All of these are covered in the relevant chapters in this book.

The luminance signal

The *luminance* signal, usually abbreviated *luma* and represented by the symbol Y, is the black and white information required to satisfy the rods in the eye. It contains information relating to brightness and contrast changes. A monochrome camera outputs a luma signal, plus sync pulses. A colour camera produces the luma component by adding red, green and blue in the correct proportions.

It can be shown that one unit of white light is made up from proportions of red, green and blue light following the expression:

$$1Y = 0.3R + 0.59G + 0.11B$$

The signal processing stage in the colour camera employs a matrix which adds the R, G and B signals in these proportions to produce a luma component.

A common test pattern used in television is the eight-bar colour display. The order of the colours is such that when the colour is removed, the bars appear with white on the left and black on the right, and descending order of grey in between. This is termed the *grey scale*. The luminance signal required at a CRT cathode to produce this display is termed the *staircase waveshape*, and a quick look at Figure 5.9 reveals how the name was derived. Bear in mind that a single staircase produces only one TV line,

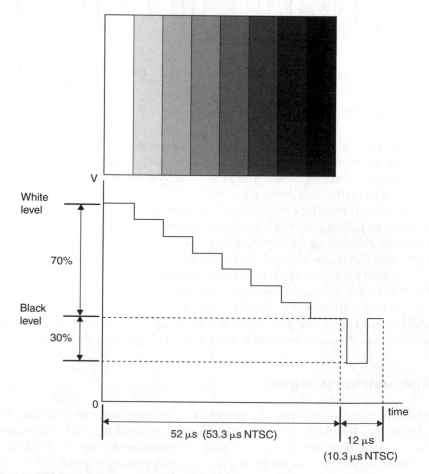

Figure 5.9 *Staircase luma signal which, when applied to a display device, produces eight grey scale bars*

and to produce a complete frame a total of 585 of these, plus 40 black lines for field flyback blanking, are required.

The CCIR standard for transmission of luminance signals along a cable requires that the voltage level should be 1 Vpp when measured from the peak white to sync tip levels, and when the input/output impedances of the equipment are 75 Ω. In this 1 V signal, the picture content will be no more than 0.7 V (70% of the total signal level), and the sync pulse will be a constant 0.3 V (30% of the total signal level).

The chrominance signal

The chrominance, or *chroma* (C), signal containing the colour information is far more complex than the luma component, and although it is not necessary for the CCTV engineer to be fully conversant with the theory, there are some essential features which he must be aware of.

The colour system employed in the UK is known as the *PAL system*, PAL being the abbreviation for phase alternating line. PAL evolved out of the earlier American NTSC (National Television System Committee) colour television system, which is used throughout the USA, Canada, Japan and parts of South America. A forerunner to PAL is the French SECAM (sequential couleur avec memoire) system; however, this is not widely used throughout the world. None of these three television systems is directly compatible and engineers must be aware of this when they are confronted with equipment that has switches (either mechanical or menu-driven) which allow different systems to be selected. For example, a DVR switched to operate in NTSC mode will record in black and white with synchronization errors if it is connected to a system comprising PAL cameras and control equipment.

In order to produce a colour picture at the monitor, it is not necessary to transmit all three primary colours. This is because the Y signal is already being transmitted, and remembering that $1Y = 0.3R + 0.59G + 0.11B$, we can see that if we transmit any two of the primary colours, the third can be recovered by matrixing the other two and the Y signal. For example, if we do not transmit the green, a matrix circuit in the monitor can be made to perform the function $G = 1Y - 0.3R - 0.11B$. Any of the three colour signals can be omitted, but green was chosen because, after it has been processed in the PAL colour encoder, it is the smallest of the three colour signals and is thus more prone to being swamped by noise during the transmission process.

To aid transmission of the red and blue colour signals it is necessary to reduce their amplitude. Because the luma is sent independently of the colour, it is possible to achieve this reduction by removing the luma content from the red and blue signals, producing *colour difference signals*, R-Y and B-Y. However, further attenuation of the colour difference signals is necessary, and thus the camera applies a weighting factor to each of them. The weighted signals are referred to as *u* and *v*. The u signal contains the B-Y information, and the v signal contains the R-Y information.

In order to transmit the *u* and *v* signals they must be modulated onto a carrier. Modulation is a process where two signals are added together in a certain way to enable them to be transmitted either through space, or along a cable. One signal is the desired information, and the other is a high-frequency carrier which is used to 'transport' the information to the receiving equipment, after which it is dispensed with. The principle is illustrated in Figure 5.10.

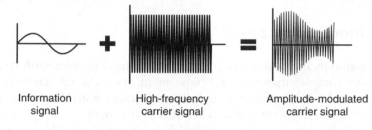

| Information
signal | High-frequency
carrier signal | Amplitude-modulated
carrier signal |

Figure 5.10 *Principle of amplitude modulation*

For the PAL colour system the carrier, known as the *colour subcarrier*, has a very precise frequency of 4.433 618 75 MHz, although it is usually referred to as the 4.43 MHz subcarrier. Similarly, for NTSC, the 3.579 545 MHz colour subcarrier may be referred to as the 3.58 MHz subcarrier. These exacting figures were chosen to ensure that the colour signal would offer minimum interference with the high-frequency luma components because, as we can see in Figure 5.11, the colour subcarrier is positioned within the luma passband.

The name PAL was given to the system because of the fact that the *v* signal changes its phase by 180° at the end of every TV line. This is unique to the PAL system and is employed to provide a built-in correction for phase errors which occur in the chroma signal as it passes through the transmission medium. Such phase errors would result in noticeable colour errors.

A colour monitor or VCR/DVR, for reasons which are beyond the scope of this textbook, uses a crystal oscillator to generate a precise 4.43 MHz (3.58 MHz NTSC) subcarrier. However, this oscillator must not only produce an accurate frequency; its output must be in exactly the same phase as the subcarrier coming from the camera. To achieve this a *chroma burst* signal is generated in the camera. This burst comprises ten cycles of the subcarrier and is placed onto the back porch of each line sync pulse. The chroma burst can be considered to be a sync pulse for the colour processing circuits in the monitor or other processing equipment and, if for any reason it is lost, the decoders in all equipment will default to black and white mode of operation.

If a camera were producing a picture of the standard eight-bar colour display, then the chroma signal, when viewed on an oscilloscope, would appear something like that shown in Figure 5.12. During the periods of the white and black bars there is no colour signal. Between the yellow and

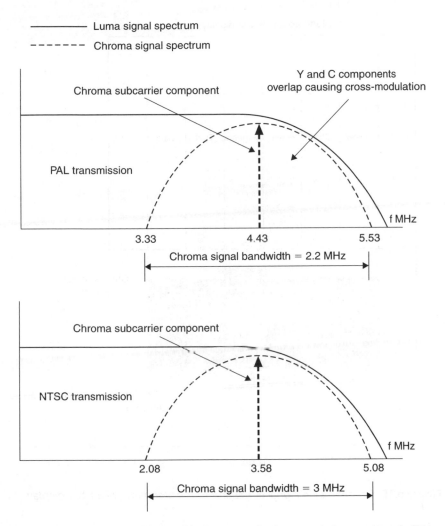

Figure 5.11 *Frequency relationship between the luma and chroma signals (PAL and NTSC transmissions)*

blue bars the amplitude-modulated subcarrier is present. Note also the chroma burst signal on the back porch of the line sync period.

Television signals

When it comes to sending the luma and chroma signals from the camera to the monitor there are a number of options available. In the CCTV industry, perhaps still the most common method is to use *composite video*, where the luma and chroma are sent simultaneously along the same co-axial cable. The prime advantage of this is the low cost compared with the other

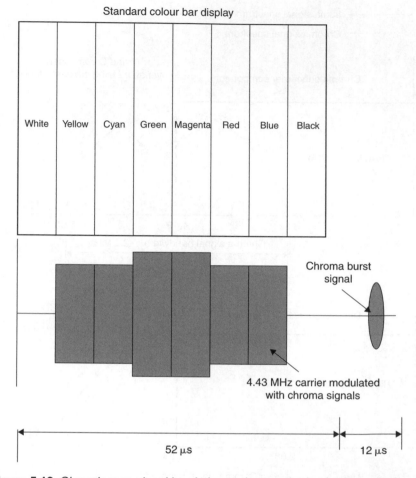

Figure 5.12 *Chrominance signal in relation to the standard colour bar display (PAL system)*

options available, because only one co-axial cable is required from each camera, and between each item of equipment in the control room.

When the two signals are mixed, the chroma signal tends to sit on the luma voltage waveshape. If we add the chroma signal shown in Figure 5.12 to the luma staircase shown in Figure 5.9, then we have a composite video signal. This addition is illustrated in Figure 5.13, where the chroma signals for the six coloured bars sit on the corresponding luma steps. Note the chroma burst signal positioned on the back porch of the horizontal sync signal. This figure also raises another point: it was stated earlier that the CCIR standard for video signals is 1 Vpp into 75 Ω, yet when a colour signal is viewed on an oscilloscope, the signal often appears higher than this, even though it is correct. The reason for this is that the CCIR standard relates to the level of the monochrome (luminance) signal which, as we saw from Figure 5.9, is measured between the sync tip and peak white

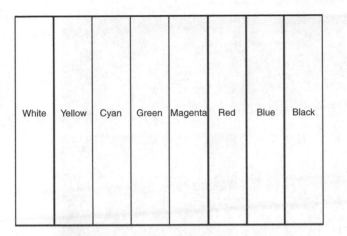

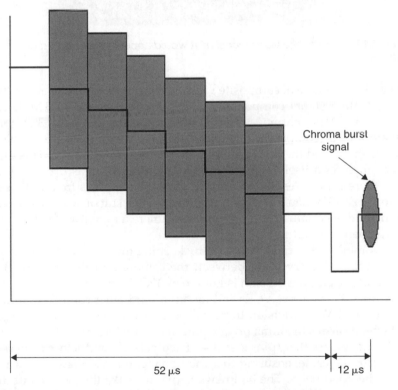

Chroma burst signal

52 μs

12 μs

Figure 5.13a *Composite video signal where the modulated chroma carrier sits on top of the luma signal*

levels. When the chrominance signal is added, the blue and red components frequently exceed the sync tip level. In conclusion, when using an oscilloscope to take an exacting measurement of a video signal waveform, a monochrome signal should be used. Alternatively, if a colour bar display is to be used, be prepared to accept a signal level in the order of 1.8 Vpp.

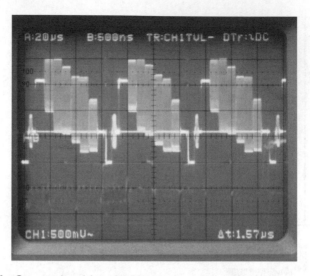

Figure 5.13b *Composite video signal as it would appear when viewed on an oscilloscope*

The disadvantage of composite video is the *cross-modulation* effect. This is where luma signal components at frequencies around 4.43 MHz mix with the 4.43 MHz chroma signal components (3.58 MHz NTSC). Once they are mixed (composite), analogue VCRs and monitors are unable to separate them, and the displayed picture contains interference. This interference manifests itself as a coloured patterning in areas of the picture where there is high-resolution luma. A classic example is frequently seen on broadcast TV when a person appears wearing a suit or jacket with a fine pattern, and at certain distances from the camera a rainbow effect appears over the person's clothing.

Cross-modulation can be avoided by keeping the luma and chroma signals separate at every point between the camera and the monitor. This method of signal transmission is known as Y/C.

Many CCTV cameras, VCRs and monitors, and other control equipment have optional Y/C provision through the four-pin S-VHS socket (Figure 5.14) which carries the luma on one pair of conductors (pins 1 and 3), and the chroma on another (pins 2 and 4). Each pair of conductors is an individual co-axial cable, ensuring that the luma/chroma signals do not mix. This is the good news. The bad news is that to make the system effective every camera requires two co-axial cables, and to further complicate things, adaptors are required to marry the large BNC connectors to the very small Y/C connectors, which in reality were never intended for use in CCTV system building. In practice it is generally considered to be too expensive and impractical to use S-VHS in a complete CCTV system installation, and use of these connectors is confined to such things as VCR input/output, etc.

In CCTV where picture resolution is of paramount importance, the losses incurred by cross-modulation are unwelcome; however, there is no

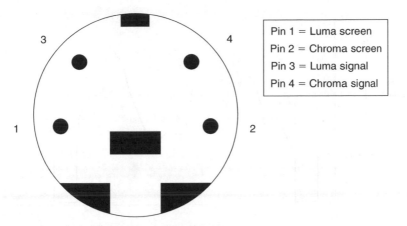

Figure 5.14 *S-VHS connector pin configuration*

cost-effective way of avoiding them in analogue transmission systems. Where analogue signals are later digitized, there are methods of detecting and largely cancelling cross-modulation effects at the control room, but this equipment is still expensive and is only feasible for larger installations. On the other hand, where the video signal remains in the digital domain from camera to control room, and S-VHS connectors are employed for monitor connection, the story is somewhat different.

Digital video signals

The concept of converting analogue signals into digital form is not new. In fact, like many of the principles employed in modern computing, the complex mathematics relating to the problem were resolved many years before the technology was available to build equipment capable of performing the operations.

We are all now familiar with digital audio, which has been around for a few decades in the form of various tape formats, and for many years in the form of compact disk. Yet digital video took much longer to make an appearance. There is a good reason for this. A digitized video signal amounts to many times more data than an audio signal of any equivalent time period, and even with the high-capacity hard drives of today's computers, without the application of the compression techniques used with digital video, an 80 Gbyte hard drive would store no more than a few minutes of video information. It was the development of video compression techniques that made digital video possible, but it has taken a number of years for this compression to be perfected, and for chip sets to be developed and manufactured in sufficient quantities to make them commercially viable.

The process of analogue-to-digital conversion is very complex, but a simple overview is shown in Figure 5.15.

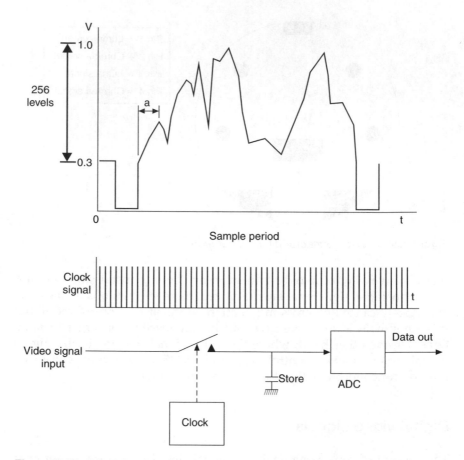

Figure 5.15 *Principle of analogue-to-digital conversion*

The video signal is fed into the circuit and the switch closes for a brief moment every time a clock pulse is present. For this illustration, we must assume that the capacitor is capable of instantly charging to the instantaneous voltage level of the video signal. When the first clock pulse arrives, the switch closes and the capacitor acquires a charge. In the brief period between the first and second clock pulses, the *analogue-to-digital* (A/D) *converter* measures the potential on the capacitor, and assigns an eight-bit binary word which corresponds to this level. When the second clock pulse arrives, a new voltage level corresponding to the video signal level at that instant is stored in the capacitor. The A/D converter now assigns an eight-bit word for this level, and the cycle repeats.

An eight-bit binary word has 256 combinations between 00 000 000 and 11 111 111, meaning that it is possible for the video signal to have 256 voltage levels. The A/D converter functions by measuring the voltage in the store capacitor, and assigning the binary word that corresponds to the closest of these 256 voltage values. Note that these levels do not include the sync components because the A/D converter only samples the 0.7 Vpp video

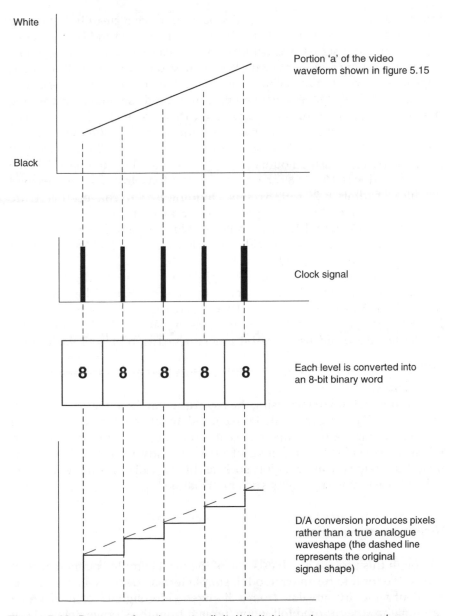

White

Black

Portion 'a' of the video waveform shown in figure 5.15

Clock signal

Each level is converted into an 8-bit binary word

8 8 8 8 8

D/A conversion produces pixels rather than a true analogue waveshape (the dashed line represents the original signal shape)

Figure 5.16 *Process of analogue to digital/digital to analogue conversion*

signal components. The sync pulses have a constant level, frequency and duration and therefore may be represented by a simple data string.

Of course, for a television monitor or VCR to process the signal, it must first be converted from its digital format back into an analogue waveform. This is known as *digital-to-analogue* (D/A) *conversion*. It is at this point where the problems associated with digitizing video signals become apparent. Consider the portion of video signal shown in Figure 5.16. If connected into a monitor, the analogue signal would produce a steadily lightening grey image along part of a TV line. However, after A/D and D/A processing, because the signal was sampled at the clock rate, the linear video waveform is now a series of steps. When viewed on a monitor, this would appear as a series of tiny rectangular blocks of increasing brightness.

The steps, or brightness increments, can be made smaller by increasing the clock rate, but this means that we produce more bytes per TV line. For argument's sake, let's suppose that we decide to produce the 780 pixels we looked at in Figure 5.4. This means that for each active TV line we will have 780 bytes, which amounts to 6240 data bits. Thus, in one TV frame we will produce 6240 × 585 active lines = 3.7 Mbits. In one second we will produce 3.7 Mbits × 25 = 92.5 Mbits. Dividing by 8 to convert this figure to bytes, we have 11.6 Mbytes per second! So we can now see that if we have an 80 Gbyte hard drive, we would be able to store 80 G ÷ 11.6 M = 115 minutes of black and white video information. A colour signal could reduce this time by up to 50%.

The figures above are just an example, but they clearly illustrate how much data is produced once a video signal is digitized. The 'recording time' of our 80-Gbyte hard disk can be increased in a number of ways. For example, we do not actually require the signal to be broken down into 256 levels. We could use a six-bit binary word to represent each video signal level. This would still give us 64 levels of grey scale, which produces a reasonably acceptable image, with an increase in the storage time to 155 minutes.

Another method of increasing the recording time is to reduce the actual area of the TV picture frame that is recorded. In other words, do not record the full TV frame area. At first glance this may appear somewhat odd, but when we consider that the field of view of many fixed CCTV cameras includes irrelevant information such as brick walls or sky, why waste valuable disk space recording such information?

Video compression

Although this process involves some of the most complex mathematics, it is possible for it to be understood in simple terms. Ideally, compression is all about removing any data from a digitized video signal which we know can somehow be restored following 'replay' from the storage device. I say 'ideally' because in practice this is often not the case. In order to achieve longer storage times on digital recording equipment, high levels of compression are frequently applied – at the expense of replay picture quality because it becomes impossible for the decoder to recover all of the discarded data.

There are a number of ways of recovering data (and therefore not having to transmit/store it in the first place). One of these is to employ different forms of *redundancy*. *Spatial* redundancy removes repeated information within a frame, *temporal* redundancy removes repeated information over a number of frames, *chromatic* redundancy removes colour information that is repeated, and *perceptual* redundancy removes information that may be considered not relevant to the image (i.e., the viewer won't miss it).

To illustrate temporal redundancy, consider the television picture in Figure 5.17a. This has one thing in common with almost all other TV pictures and that is, between one TV frame and the next (1/25th of a second), most of the picture information does not alter. Thus, having digitized all of the picture once and stored it, what is the point in storing the same data again and again? Obviously we are referring to the background areas which for most of the time are stationary; that is, until either the camera pans, or the picture cuts to another shot.

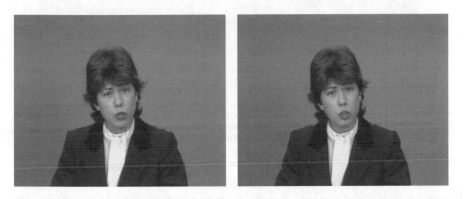

Figure 5.17a *A typical television image contains a lot of redundant information. In this example of two consecutive TV frames there are subtle changes in the face, and the body has moved very slightly*

Figure 5.17b clearly illustrates the effectiveness of temporal redundancy by showing only the difference information between one frame and the following frame. From this illustration we begin to appreciate just how much data can be excluded in transmission/storage of subsequent frames without any loss of information in the final reconstituted picture.

So, one way of compressing a digital video signal is to remove duplicated data, thus making it redundant. To rebuild the picture, the compression decoder simply uses the same data over and over again to produce the walls and floor in our example. Another form of compression is *prediction*. Looking again at the example in Figure 5.17, when the camera pans, all of the repeated data will move in the same direction. Upon seeing this, the compression processing chip, instead of passing all of the wall and floor area for storage, produces a relatively small amount of mathematical data which, during the recovery process, will be used to predict where the wall and floor will have moved to on a frame-by-frame basis. In other words, the same data is used over and over just like before, only now it is being moved around the screen.

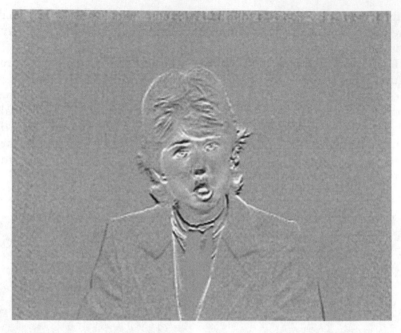

Figure 5.17b *Illustrating temporal redundancy. Here only the changes between two consecutive frames are shown. (Illustrations courtesy of Tim Morris)*

As outlined at the start of this section, we cannot expect to heavily compress a signal and not lose something in terms of quality. The compression chip sets are so designed that equipment manufacturers using them can set the level of compression, and the rule is very simple. A lot of compression results in a lot of redundant data, which means long recording times, but at the expense of picture quality. Minimal compression means that a lot of data need to be stored, reducing the recording time, but producing high-quality (resolution) pictures.

Compression is measured as a ratio of the amount of data entering a compression system to the amount that comes out. Thus, a compression ratio of 1:1 indicates no compression, 5:1 would indicate that there is five times less data at the output, 10:1 indicates a reduction of ten times, and so on.

The process of decompressing the data involves taking the data that has been stored and applying it to complex mathematical algorithms. The success with which the original signal is restored depends upon the amount of original data that was retained (that is, the amount of compression applied) and the effectiveness of the algorithms (that is, how well they have been derived). The term 'lossless compression' is becoming used more and more, but only where little compression has been applied in the first place. At the time of writing, once even moderate amounts of compression have been applied to an image, some information loss and/or artefacts will be evident, even if only on certain types of picture information. This is because a point is reached where the decoder has to make a 'best guess' at what information should be put back into a certain picture location and each time it gets this wrong, an unwanted pixel is produced on the picture (i.e., an artefact).

It is difficult to relate a specific compression ratio to any given number of artefacts because there are too many variables. For any given compression ratio, the point at which compression artefacts become noticeable depends on the picture information, the type of compression used, the algorithms employed by the manufacturer and the processing circuitry. Nevertheless, as an aid to installers and end users, manufacturers often compare a given compression ratio with known analogue formats such as VHS and S-VHS. In practice it is difficult to make direct comparisons because the artefacts produced by digital compression errors often manifest themselves quite differently to those produced by analogue processes, but making such comparisons does serve the purpose of giving some idea as to how the reproduced image will appear.

MPEG-2 compression

Digital CCTV equipment employs one of two forms of compression: MPEG or Wavelet.

MPEG (Moving Pictures Experts Group – a body set up by the ISO in 1988 to devise standards for audio and video compression) video signal compression is a very effective, robust and reliable compression format. MPEG-2 is a natural evolution from MPEG (or MPEG-1) and, although it is the compression format used in a lot of CCTV equipment, it was primarily developed with broadcast television and home entertainment in mind.

A TV signal can be said to be four-dimensional, having attributes of sample, horizontal axis, vertical axis and time axis. All of these can be explored for redundant information by a compression encoder, and MPEG-2 employs three types of compression to explore these four attributes.

The first type of compression is temporal redundancy, also known as *inter-frame* (P frame) compression because it is applied across the frames. This type of compression takes account of the fact that much of the video information is the same from one frame to the next and therefore it is only necessary to transmit the differences between the frames. Consider the image in Figure 5.18. In this case, assuming that the camera angle is fixed, the only significant changes between frames will be the movements of the vehicle in the centre of the picture. Thus we see that throughout the entire scene when the vehicle is moving round the corner, which for the purpose of this example we shall assume lasts for three seconds, large amounts of data in the 75 frames will be identical. If only data relating to changes in picture information are transmitted, it may be possible to reduce the data content by 90%.

The second type of compression is spatial redundancy, also known as *intra-frame* (I frame) compression because it is only applied within a single frame. Looking at the image shown in Figure 5.19, assuming a horizontal

Figure 5.18 *The only significant movement in this clip is that of the vehicle, and therefore the majority of the video information only needs to be transmitted for the first frame*

Line 10

Figure 5.19

resolution of 720 TVL, during each active line period many of the 720 luma samples, or pixels, will be identical. Therefore, it can be reasoned that it is only necessary to transmit the first pixel in a sequence, plus information to inform the decoder how many times the pixel must be repeated. For example, if we consider line 10 of the image in Figure 5.18, we can see that instead of sending 720 pixels of data at 8 bits per pixel, we need only send the 8-bit value of pixel number 1, along with a small amount of data relating to the position (line 10) and the number of repetitions (719).

Intra-frame compression also makes use of the fact that, in general, the amplitude of video components reduces as their frequency increases. Therefore, if we assign fewer bits to the high-frequency end of the spectrum, a further saving can be made. However, this *will* lead to increased noise in areas of picture that contain a lot of fine detail but, as long as the process is not pressed too far, the eye will not easily discern this.

The third type of compression is *statistical redundancy*, also known as *prediction or motion compensation*. Looking again at the image in Figure 5.18, once the vehicle is fully in shot, it is a relatively simple operation for the encoder to produce a plot of the pixel movements across the 75 frames of the video clip. Therefore it is not necessary to transmit all of the data relating to the changes between frames, but simply the prediction data required by the decoder to move the pixels between each frame.

Further data reduction may be introduced by replacing the regular and predictable line and field sync signals with short codes that will indicate to the decoder the start of these periods. Similarly, short codes may be used to indicate black levels, which occur frequently in most video signals. And finally, replacing long repetitions of '0' or '1' with shorter words that simply indicate the number of repetitions can reduce the amount of data considerably.

Figure 5.20 shows how MPEG arranges the TV frames for storage and subsequent transmission to the decoder. This illustration shows how one I frame is transmitted on every 12th frame to serve as a 'detailed' reference block, with P frames (predicted frames) interleaved between I and B frames.

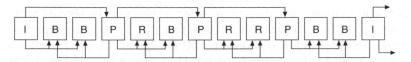

Figure 5.20 *The MPEG frame structure comprising I, P and B frames*

The B frames (bidirectional frames) are made up purely from interpolated information from adjacent P and I frames. One weakness with this system is that an error in an I frame will propagate through the following frame sequence until the next I frame. Such an error would persist for about 0.5 s. Errors occurring later in the sequence will endure for a lesser time period. The twelve-frame cycle illustrated in Figure 5.20 relates to a typical broadcast MPEG-2 transmission although, in practice, the encoder designers may employ more or fewer I frames, depending of the application and/or amount of compression desired.

At this point we may be forgiven for believing that we have reduced the data by as much as possible; however, it is here that the mathematicians take over. The remaining 8-bit samples are arrayed into 8×8 matrices where a process called *discrete cosine transformation* (DCT) is performed. This is followed by a *quantization process*, resulting in a large reduction in data per frame compared with the original 8-bit samples. DCT is a key part of MPEG because this part of the compression process is truly lossless as every bit can be recovered at the decoder by reversing the mathematical process. On the other hand, quantization is distinctly lossy, and these losses are irreversible.

Within the encoding and decoding process there are a number of errors that may occur. Data errors can occur even though error correction is employed in MPEG because, if an error is too great, entire 'slices' of picture will be rejected because of the way in which data is arrayed within a frame. Decompression errors, although not an error as such, may occur because of excessive removal in the redundancy process (i.e., a high compression ratio). Where large amounts of information have been identified as redundant there is no way that the original data (and thus image) can be restored. This results in a 'blocky' effect (pixilation) over areas of the picture. Finally, data may be insufficient because of 'overloading' at the encoder. Overloading describes a situation where the video clip contains a large amount of high-definition, fast-changing information. In such cases, each frame bears little resemblance to the adjacent frames in the footage and therefore there is little, if any, redundancy. Consequently the encoder is unable to output the large amount of data necessary to communicate the images, resulting in large areas of pixilation and/or numerous freeze frames. In a CCTV system, this situation is frequently encountered whilst a camera is panning rapidly.

MPEG-4 compression

Available in October 1998, this compression format may be seen as MPEG-2 with a lot of additional functionality and features. The basic compression/decompression remains the same as for MPEG-2; that is, it still employs spatial and statistical redundancy, 8×8 pixel blocks and DCT. Where MPEG-4 differs is in the functionality options available to system and equipment designers.

It was stated earlier that MPEG-2 was developed primarily with the broadcast and home entertainment industries in mind, i.e. digital television, surround sound, DVD, etc. However, it must remembered that these are all two-dimensional video formats where all that is required is a high-resolution picture with good-quality sound. By the early 1990s it was becoming clear that the three emerging technologies of commercial film/TV/entertainment, computing and communications (in particular, the Internet and network communications) were rapidly converging, and that a common standard for compression/decompression of this multimedia video and audio content would be required.

To meet this requirement, MPEG set up the MPEG-4 committee, which was given the task of producing a suitable standard. They first met in September 1993. As this committee analysed the problems, they realized that a single set of compression rules set in stone (like MPEG-2) would never be able to satisfy the complex and varied requirements of the converging technologies. What they eventually devised was a whole range of digital video/audio manipulation tools that would all be included in the MPEG-4 standard, and from which system and equipment designers could draw. Consequently, different items of equipment which incorporate MPEG-4 and perform the same function may in practice utilize a different tool set to achieve their objective. From the point of view of CCTV equipment, many of the available tools and functions are not really relevant as they belong in the world of multimedia production and delivery. Nevertheless, some CCTV equipment does utilize MPEG-4 and therefore it is helpful for the CCTV engineer to have an overall idea of MPEG-4.

So what does MPEG-4 offer in addition to MPEG-2 capabilities? Well, to begin with, MPEG-4 has the ability to texture map 3D as well as 2D objects using a Wavelet algorithm (Wavelet will be discussed later in this chapter) although, at the time of writing, no one has actually managed to develop a method of identifying these objects in real time. Another significant feature is its robustness, even when the bit error rate is high, making MPEG-4 an ideal choice for digital video signal transmission over Ethernet and wireless LAN.

However perhaps the most significant feature is the ability of MPEG-4 to separately identify objects within a scene, which enables these objects to be individually manipulated and also interacted with by the user. Figure 5.21 illustrates the primary video analysis methods used by MPEG-4 to separate video information in order to sample, compress and manipulate the information in a video sequence with maximum efficiency. Because objects are identified as separate entities (2D background, arbitrary shapes, 3D, etc.), the encoder is able deal with them as such from the point of view of both compression and manipulation. In Figure 5.21, the main point of interest is the goalkeeper, although the two girls in the background also represent areas of lesser significance. In MPEG-4 these three objects can be identified as arbitrary shapes and processed separately from the rest of the video image. Although MPEG-4 utilizes 8×8 pixel blocks and applies DCT to each block as in MPEG-2, it also has a feature known as *shape-adaptive DCT* which is applied to arbitrary shapes such as the three girls in our illustration, with the effect of improving shape accuracy and coding efficiency.

It is also possible to apply differing levels of compression to different arbitrary shapes. For example, objects of little interest could be subjected to higher levels of compression than those of greater interest, thus improving the S/N ratio and image quality in the areas of greater interest. Referring again to our image in Figure 5.21, the girl in the foreground is clearly more significant than the two girls in the background and therefore these two objects could be subjected to a higher level of compression than the data relating to the goalkeeper.

Arbitrary shapes do not have to come from the original video source. They could in practice be computer-generated objects (animations, etc.), or video objects that have been extracted from another video source. Such

Arbitrary shapes
of lesser interest

Arbitrary shape deemed to
be of high interest; potential 3D object

Figure 5.21 *A TV image may be broken down into component parts by the MPEG-4 encoder in order that each part may be processed separately. This technique greatly improves the efficiency of the compression process as well as offering features such as user interaction and the ability to add images from other sources. (Photos courtesy of David Close).*

objects may be inserted into each frame in a video sequence, even though they are a sequence in their own right.

Figure 5.22 illustrates how MPEG-4 breaks an image down into separate component parts. The main background forms the largest (2D) object and is known as a *sprite*. Assuming that the camera is stationary, the sprite contains all of the data that need only be transmitted occasionally (I frames) and for all other frames can be replaced by a single 8-bit word which informs the decoder that it is to use the same sprite again. When we consider that one 8-bit word has replaced thousands of pixels, we can immediately appreciate how much compression has been applied. The sprite may be much wider than the display area; for example, the camera may be fitted with a fisheye lens that produces a 360° image, but we only wish to view a part of that image (with all distortion removed) on a monitor.

For most CCTV applications the goal netting in the image would probably be included in the sprite. However, for multimedia applications it could be separately identified and encoded as a 3D object, thus bringing it out to the forefront of the image.

MPEG-4 is capable of defining the moving parts in the image as separate, arbitrary shapes which can be both compressed and transmitted separately. As discussed earlier, this means that the encoder can apply different levels of compression to each object, depending on the perceived importance of the object, and can also apply different animation tools depending on the type of object (video, synthetic animation, etc.). This makes for maximum compression efficiency, optimum S/N ratio for objects of greater interest,

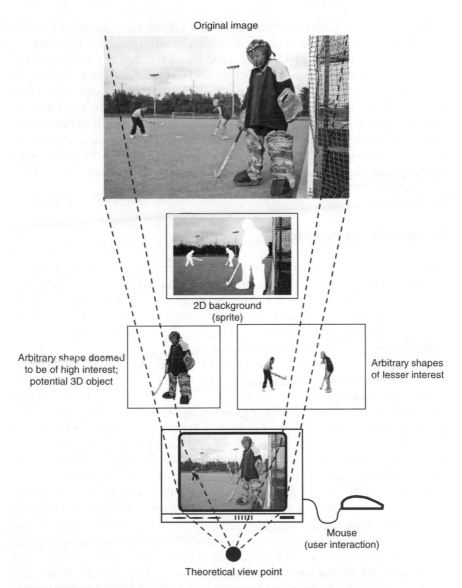

Original image

2D background
(sprite)

Arbitrary shape deemed
to be of high interest;
potential 3D object

Arbitrary shapes
of lesser interest

Mouse
(user interaction)

Theoretical view point

Figure 5.22 *Illustration showing how MPEG-4 separates images into components. These may then be transmitted separately and re-constructed by the decoder. Where the sprite is much wider than the viewing area, the encoder must be able to determine the desired viewpoint in order to crop the image accordingly. (Photos courtesy of David Close)*

improved ability to round the edges of objects (by use of shape-adaptive DCT), and simplified motion prediction. The decoder receives the separate images (plus audio, if present) and, making use of the sophisticated and robust synchronization tools, re-constructs the original image.

Finally, MPEG-4 incorporates tools that enable user interaction with objects, which is an essential feature in many multimedia applications.

From this discussion of MPEG-4 it can be seen that it is a truly remarkable toolkit for multimedia production and data transmission. However, at the time of writing, MPEG-2 remains the more popular of the two formats for CCTV multiplexer and DVR applications where the requirement remains, primarily, a two-dimensional video image, usually without any sound. Nevertheless, IP cameras find MPEG-4 a more suitable format, as do the emerging range of 360° cameras that incorporate high levels of digital processing in order to produce a number of distortion-free 4:3 TV images from the one 360° sprite.

Wavelet compression

Signal analysis using wavelets is based on Fourier analysis, which was formulated by Joseph Fourier in the 1800s, although it was not until the 1980s that wavelet analysis was used as a method of signal analysis. This type of analysis has many applications including astronomy, music, earthquake prediction, radar, neurophysiology and – of course – video signal compression.

Wavelet analysis does not break the TV frames down into blocks as in MPEG; rather, it analyses each frame individually and as a whole. In other words, there are no 'P' or 'B' frames. Figure 5.23 illustrates how wavelet analysis separates the signal components in each frame into frequency bands (typically 42 although, for simplicity, this illustration only shows 10 discrete bands). Bearing in mind that each frequency band represents video components of a particular resolution, from this illustration it is possible to identify the picture content contained within each band.

Having identified the components within a TV frame, it is now possible to individually analyse each band and decide, from a compression point of view, which components *must* be retained to maintain image integrity and which components may definitely be discarded without detriment to the final reconstructed image. What happens to the remaining components will be determined by the level of compression set at the encoder. Generally speaking, the signal components which are deemed to produce information that is not visible to the human eye are discarded (typically picture content of fine detail). Spatial compression is then applied to the remaining bands and, finally, applied algorithms provide further compression of the remaining data. This compression is non-linear with the greater compression being applied to the higher components (Figure 5.24).

Each frame is processed individually in real time, and is scanned three times at the encoder to determine the optimum compression ratio for that frame. Compared with MPEG, wavelet encoding is very much simpler and therefore less expensive, a factor that can make Wavelet more attractive for CCTV applications. In broadcast television there is only one (MPEG)

Figure 5.23 *How Wavelet sees an image. The original image is broken down into discrete layers in terms of picture resolution. Different levels of compression can then be applied to each individual layer. (Photo courtesy of Tim Morris)*

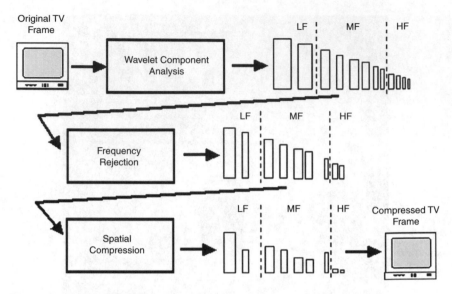

Figure 5.24 *Simplified illustration of the wavelet compression process*

encoder – at the transmitter – whereas in CCTV we have to build an encoder into every multiplexer, digital recorder, etc. This makes the simpler wavelet encoder more attractive from a manufacturing point of view, although large-scale integration in silicon chips is tending to close this gap.

Because each frame is individually processed, Wavelet is much simpler to edit. In CCTV, fast-switching multiplexers can result in MPEG having to create a lot of I frames, thus increasing the data capacity. This is not the case with wavelet compression.

Wavelet technology allows much higher compression ratios than MPEG for the same image quality (or resolution). A wavelet compression ratio of 20:1 tends to be comparable to that of an MPEG compression ratio of 10:1, although the losses in both systems are quite different, making a direct comparison difficult. This being the case, the larger data file size of wavelet compression is countered by the fact that greater compression ratios can be applied.

Because it does not break the picture down into the 8 × 8 blocks, wavelet compression does not normally produce the 'blocky' effect associated with MPEG when larger amounts of compression are applied. Rather, a heavily compressed wavelet image takes on a blurred appearance, perhaps comparable to looking at a picture produced by a camera with a slightly defocused lens. Having said this, it is possible to experience a blocky effect with wavelet compression if the encoder design utilizes particular (usually simpler) *basis functions*. In Wavelet algorithms there are numerous basis functions, and the encoder designer will select certain ones, depending on the image quality that they wish to achieve and the computational complexity they are willing to afford.

In conclusion, wavelet compression sees the TV frame as a number of layers of differing resolution. Each layer is individually analysed to determine whether it needs to be retained, and those that are retained are further analysed to determine how much (if any) compression may be applied. Higher compression levels are applied to the higher-resolution layers – however, this amount is variable. As with any compression system that is based on mathematical algorithms, there are many ways that the compression may be applied, with varying results. The problem with wavelet compression throughout the 1990s was the lack of any industry standard and therefore, in the year 2000, the Joint Photographic Experts Group (JPEG) released the *JPEG-2000* standard, which is the official standard for wavelet video signal compression.

Common interchange format (CIF)

Sometimes referred to as the common intermediate format, these standards specify picture formats in pixel sizes. CIF itself describes an image size of 352 horizontal × 288 vertical pixels (352 × 240 NTSC) before compression, which amounts to one quarter of the number of pixels in a broadcast TV picture frame. 2CIF describes an image that contains twice the number of pixels as CIF, which is 704 × 288 (704 × 240 NTSC) and 4CIF describes an image containing four times CIF: 704 × 576 pixels (704 × 480 NTSC). Note that for 2CIF, only the horizontal resolution is increased as this in effect doubles the number of pixels, and 4CIF doubles both the horizontal and vertical pixels, making it equal in resolution to a full broadcast frame.

Other CIF specifications include *QCIF (quarter CIF)*, which defines a resolution of 176 × 144 pixels, *SQCIF (sub-quarter CIF)*, which defines a resolution of 128 × 96 pixels, and 16CIF, which is 1408 × 1152.

ITU-T recommendations

Over the years the *ITU* (International Telecom Union) has produced a number of standards which deal largely with two-way audio/video communication. The ITU-T *H.320* standard is a recommendation that specifies the requirements for low-bandwidth audio/visual telephone and video conferencing equipment and services. As such, H.320 comprises a number of other (H.26×) substandards.

Published in 1990, the CCITT *H.261* standard is the substandard that defines video compression for H.320. The specification was defined with video transmission over ISDN (Integrated Services Digital Network) in mind and, as such, specifies data rates in the order of 64 kb/s and 128 kb/s. Images are non-interlaced and the DCT coded output comprises luminance and colour difference signals which may be either CIF or QCIF resolution.

The compression methods applied contain 8 × 8 pixel blocks, intraframe and prediction similar to those described earlier in this chapter.

Owing to its low resolution (which is important in order to maintain a low bit rate when transmitting over a low bandwidth channel such as ISDN), H.261 does not perform well in CCTV applications.

A more effective specification is the *H.263*, which is a natural evolution of H.261. This standard specifies resolutions that are higher than H.261 but at lower bit rates, which is made possible by more advanced compression techniques being applied. The output may be SQCIF, QCIF, CIF, 4CIF, or 16CIF. The standard was introduced during the mid-1990s to accommodate video data transmission over the very narrow bandwidth communication channels provided via PSTN and GSM modems, which are typically 9600 bps.

A more recent standard (May 2003) is the *H.264*, which is also specified under a different but familiar standard, *MPEG-4 Part 10*. The reason for having two identical standards is because they were developed jointly by the MPEG and the ITU-T groups, who set out to create a standard that would exceed MPEG-2, MPEG-4 Part 2 and ITU-T H.263 in terms of image quality and bit rate.

H.264 offers bit rates at half that of H.263/MPEG-2 without a comparable loss of image quality. This is made possible by the application of a range of compression tools, some of which are new, and many of which are improvements on existing tools employed in earlier compression specifications. For example, motion detection spanning many frames offers large reductions in data when video sequences contain scene cuts that jump rapidly back and forth (which is basically what happens when a CCTV DVR is recording multiple cameras simultaneously). The pixel block sizes are also no longer fixed at 8 × 8 but are dynamically variable between 16 × 16 and 4 × 4, which provides very precise definition of moving arbitrary objects and reduces aliasing, resulting in sharper images.

6 The CCTV camera

The camera can be considered to have two parts: the pick-up device(s), and the signal-processing circuits.

For many years cameras relied on thermionic (valve technology) tubes to derive a signal voltage from the incoming light information. However, cameras employed in modern CCTV systems use a solid-state pick-up device known as a *charge coupled device* (CCD).

Similarly, advances in technology have meant that the traditional analogue signal-processing techniques which have served us well for many years have given way to digital signal processing, meaning that cameras are able to produce remarkably clear pictures under very hostile lighting conditions.

Charge coupled device

The CCD is a silicon device which can store an electrical charge. A chip containing a number of CCDs in an array can be used to store samples of analogue video or audio signals where they can be manipulated. And so, although the CCD chip is not in itself a digital device, when controlled by a microprocessor it can be used to move analogue samples around in the fashion of a shift register.

The CCD design can be modified such that electrons are released when photons (light) fall onto the device. Thus the CCD behaves somewhat like a photodiode. If the light output from a lens is focused onto an array of these photodiodes, each diode will derive an output voltage proportional to the amount of light falling upon it. Thus the chip is converting light energy into proportional electrical charges.

A typical imaging chip used in a CCTV camera contains many thousands of CCDs arrayed in a rectangular pattern. As we shall see in a moment, the voltages from the cells are integrated to create individual *pixels* ('pixels' is derived from the term *picture elements*). For a CCD image device, the picture resolution is determined by the number of cells in the chip, and the density of the cells. In theory a ½" CCD chip will have better resolution than a ⅓", and this is true if we compare like for like. However, a modern ⅓" chip can have a greater cell density than an older ½" device, which means that a new ⅓" camera could possibly offer a higher resolution than the old ½" camera it is replacing. As a general rule, the greater the cell density, the higher the cost of the chip.

CCDs can have a tendency to overload under bright light conditions, causing light areas of the picture to diffuse into a white mass. This effect is termed 'burn-out'; however, this is not inferring that the CCD itself is damaged, only that the picture quality is degraded. The burnout effect can be avoided by using suitable filters and a good-quality iris control.

A typical CCD chip is shown in Figure 6.1.

Figure 6.1 *A 1/3" CCD imaging chip employed in CCTV cameras*

CCD chip operation

A *frame transfer* chip is illustrated in Figure 6.2 and, although this chip is no longer employed in TV cameras, it does serve as a good starting point when describing the shift register action of a CCD chip.

Although called a frame transfer chip, this early device actually operated at field rate. During one field period of 20 ms the imaging CCDs gather a charge from the incident light. Then, during the field flyback period, the charges are shifted downwards through the image CCD cells into the CCD storage area. During the period of the following field the image devices re-charge, whilst the information in the storage area is moved one TV line at a time into the horizontal storage area, from where it is clocked out over a 52 µs (one active TV line) period. At the end of the field period, the cycle repeats.

The problem with this method of charge transfer is that the charges have to move down through the image area to get to the store area, meaning that the image CCDs must remain active not only throughout the field period, but also during the charge transfer period. The charges are therefore 'topped up' as they travel down through the image area by the incident light which is still falling onto the chip, resulting in a vertical smear in bright areas of picture content.

The solution to the vertical smearing problem was to re-design the CCD image chip, and employ a charge transfer technique known as *interline transfer*. The chip architecture for this is illustrated in Figure 6.3.

With this architecture, each CCD charge can be moved directly into its allocated temporary store area without having to travel through the other

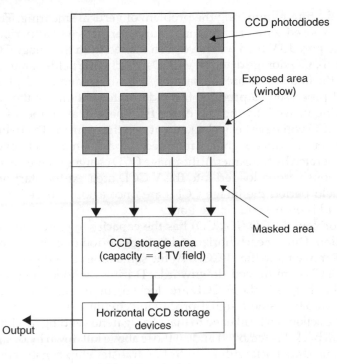

Figure 6.2 *Frame transfer CCD chip principle*

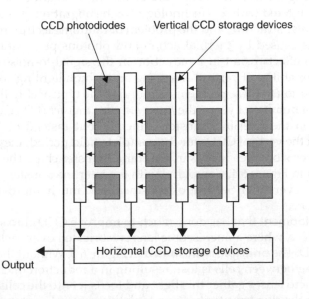

Figure 6.3 *Vertical and horizontal CCDs in an interline transfer image chip. The time that it takes to clock the information out of the H-CCD is 52 μs – one active line period*

image cells, thus eliminating the problem of vertical smearing. The image cells are exposed for a brief moment during an active field period. A previously acquired TV field will have been moved from the image CCDs into the vertical CCD storage devices (V-CCDs) during the field-blanking period. Thus, as the image cells are re-charging, this information is clocked at line rate, and processed to produce the video information for the field currently being viewed. By the end of the field period, all of the information in the V-CCD will have been clocked out and thus, during the field flyback period when no video signal is required (other than a black level which can be generated by the camera), the image CCD charges for the next field are simultaneously downloaded into the V-CCD area. At the start of the following field period the image CCDs are once again exposed, whilst the stored field is once again clocked out.

The horizontal CCD (H-CCD) has the capacity to store one TV line of information. During each horizontal flyback period one charge from each V-CCD is moved into the H-CCD. At the same time all of the other charges in the V-CCDs are moved downwards. During the following active line period, the charges in the H-CCD are clocked out in serial form. From here they pass into the camera signal-processing circuits to be output as a video signal. The action can be likened to that of a cigarette vending machine where the bottom packet is removed and all those above fall down by one position.

The main drawback with the interline transfer chip construction lay in the fact that the image surface is no longer occupied solely by image CCDs. Because the vertical CCDs are adjacent to the image devices, a lot of the light falling onto the chip is unused. This reduces the sensitivity of the CCD chip, making it less able to cope under low light conditions. Other problems encountered with this technology are the migration of charges within the molecular structure, and the problem of random electron release in the store areas caused by thermal action and photons penetrating the store area. Such effects result in a reduction in the signal-to-noise (S/N) ratio and the possibility of vertical smearing under high lighting conditions.

The solution to these problems is found in the type of chip that has been in use for a number of years, the *frame interline transfer* (FIT) chip. This still operates on the interline transfer principle, but instead of holding the charges in the vertical CCDs for the complete field period, they are moved into a lower store area just as in the frame transfer chip. The principle is illustrated in Figure 6.4. Compared with the interline transfer chip, the FIT chip offers a very high S/N ratio, low smear, and much improved low light performance.

A development that has done much to improve CCD chip sensitivity is the *micro lens*, where a microscopic lens is fabricated over each individual image CCD. The principle is illustrated in Figure 6.5. Without the micro lens, light falling between cells is lost, resulting in a reduction in chip sensitivity. The micro lenses gather this light and focus it onto the cells, effectively increasing the chip sensitivity. Sony took this principle a step further with the introduction of the *exwave* chip, which employs a more effective micro

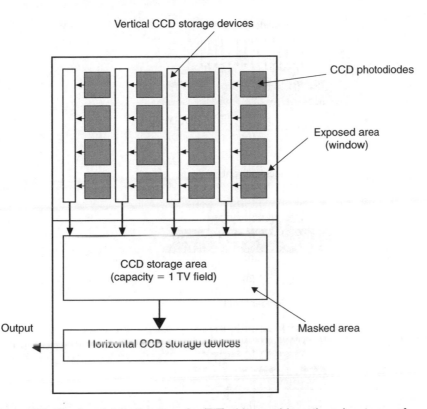

Figure 6.4 *The frame interline transfer (FIT) chip combines the advantages of the frame transfer chip (having a masked storage area safe from corruption from incident light) and the interline transfer chip (adjacent storage areas removing the effect of vertical smear)*

lens, as illustrated in Figure 6.5c. The majority of CCTV cameras used today employ Exwave technology CCD chips.

An alternative to the CCD chip technologies discussed so far is the *digital pixel system* (DPS) image sensor produced by Pixim. The main difference between this and a conventional CCD image chip is the move away from the analogue shift register principle. DPS image sensor chips have an analogue-to-digital (A/D) converter directly attached to each individual photodiode pick-up, or pixel. This means that the image chip is able to perform complex processing functions even before the information is clocked out, resulting in very high sensitivity, high resilience to burn-out, wide dynamic range, high S/N ratio and excellent colour quality.

The principle of operation is shown in Figure 6.6. The A/D converter on each photodiode is connected to a data bus where its binary output (relating to the incident light level) is passed to a RAM. Because the information is being shifted in digital form, the problems of smear and noise associated with incident light are eradicated.

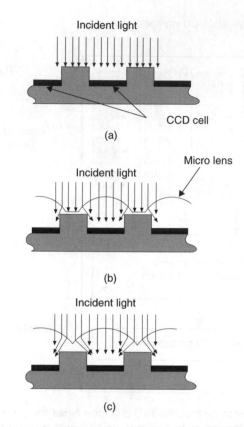

Figure 6.5 *(a) In a conventional arrangement, light falling between cells is lost. (b) The micro lens increases the light input to each cell, thus improving the sensitivity of the chip. (c) Sony Exwave technology results in even better light-gathering ability*

DPS chips offer a very wide dynamic range (that is, the range between the darkest and the brightest visible information in a scene) and have a high resilience to burnout caused by overexposure because of the unique way that each pixel samples the incident light. In a CCD chip, each cell charges up to a voltage potential that is proportional to the level of light that is falling upon it. However, this means that under low lighting conditions the charge level will be very low, making the signal voltage liable to fluctuations caused by the action of free electron charges in the chip or, in other words, background noise. At the opposite extreme, under high levels of lighting, the cells will charge to their peak and the image will appear burnt out. DPS chips do not sample the light level in the same manner. Because the charge created by the incident light is immediately converted into a binary value, the chip is able to monitor the rate of rise of the charge value at each photo-diode and, from this information, determine not only the peak level but

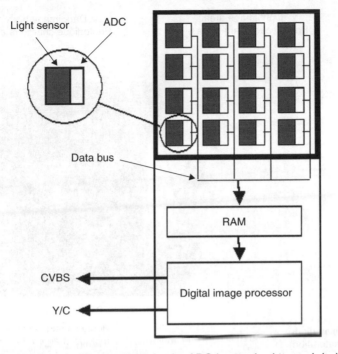

Figure 6.6 *DPS image sensor principle. An ADC is attached to each individual cell/pixel; therefore the light input can be continuously sampled, enabling a very accurate measure of the light level at each point on the pick-up area*

also the extreme light levels (i.e., very dark/very bright). With this information, the chip can decide at which point in time each pixel will take its sample, thus avoiding the problems of having some parts of the picture under/overexposed.

The principle is illustrated in Figure 6.7. The photodiodes relating to the dark areas in this photo will take a considerable time to accumulate a charge, and so the chip waits until the point in time just before these cells are about to overexpose (time t3 in Figure 6.7) before taking the sample. The digital image processor is then able to calculate the true light level at these cells by equating the rate of rise of charge over the time taken to reach a maximum value. The photodiodes relating to the light areas of the image will reach a maximum charge value much more rapidly, so the processor will use time t1 to calculate this light level. Thus it can be seen that the DPS chip avoids having to take samples at low cell charge levels (so avoiding poor S/N ratio in dark picture areas) and does not suffer from burn-out because a sample will always be taken just before each photodiode cell reaches its saturation level. In other words, DPS chips dynamically sample the image.

The dynamic range is improved using this dynamic sampling technique because information in the extreme light and dark areas of the image are not

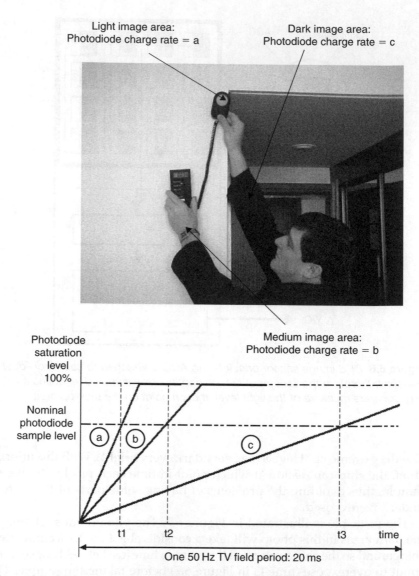

Figure 6.7 *DPS image chips sample the light level at a point where each photodiode has a high charge potential. The actual light level can then be computed from the time that it took for the charge to reach this level. This ensures a good S/N ratio and no image burn-out*

hidden by overexposure or background noise. There are, however, some viewing conditions where a high dynamic range can prove counter-productive. For example, where a scene is generally dark but has a bright area within it, the lens iris will by nature of its operation open up to accommodate the darker area. This will cause the bright light content to reflect inside the lens and scatter before reaching the image chip, resulting in an apparent misting of the view. This does not mean to say that this effect is inevitable when using cameras that employ DPS chips. The camera manufacturer can take steps in the design to restrict the dynamic range under such conditions.

Electronic iris

Interline transfer made possible the introduction of *electronic iris* (EI). This can be compared to the mechanical shutter in a photographic stills camera; however, it is performed electronically by the application of a voltage to the cells. The electronic iris circuits in the camera adjust the charge time of the CCD cells to suit the average incoming light level.

The advantage of EI is that, where a fixed iris lens is used, the camera is able to compensate for changes in lighting levels. However, where the lens has an *automatic iris* (AI), the electronic iris should be switched off to prevent an effect known as 'hunting'. This is where, following a rapid and large change in light level, both iris circuits react. However, the EI normally reacts first, and thus when the mechanical iris closes a moment later the light input becomes too low. This causes the EI to 'open' once again, followed quickly by the mechanical iris, so the light level is once again too high, and the cycle repeats a number of times until the iris circuits stabilize. This oscillation can be sustained at a very rapid rate for quite long time, resulting in a rapid changing in brightness level along each television scanning line, producing a patterning effect over the whole picture that could be mistaken for an RF interference effect.

IR filters

CCD image chips are generally sensitive to IR radiation, which is what makes them so sensitive under low light conditions. At first glance this might appear to be good news, and in some instances it may be. However, unrestricted penetration of IR radiation into a CCD image chip can cause problems.

Because of the longer wavelength, IR radiation penetrates deeper into the silicon substrate of the CCD chip than visible light, and this penetration can lead to the undesirable release of electrons in the charge storage areas, changing the values of the wanted charges, causing smearing and loss of definition. To prevent this phenomenon an *IR cut filter* is placed over the light input window of the CCD chip. This reduces the sensitivity of the chip to some degree; however, the improvement in definition and S/N ratio makes it a worthwhile trade-off.

Monochrome cameras employing chips without an IR filter are available, and find applications where the area is to be illuminated by IR spotlights, or where low light level operation is required. All colour CCD chips must have an IR filter in order to produce accurate colours.

Colour imaging

Up to this point we have only considered the operation of the monochrome image chip, which only produces a luminance signal. Producing a colour signal is somewhat more involved because the chip must be able to generate three signals: red, green and blue. There are two ways in which this can be achieved: by using three chips, or by using a single chip with a colour filter.

The three-chip method, illustrated in Figure 6.8, is by far the best. The incoming light is split into its three component parts using an array of mirrors, including special *dichroic mirrors* which reflect some frequencies of light whilst allowing other frequencies to pass through. Each CCD operates in very much the same fashion as it would in a monochrome camera, producing picture information relating to the colour of light falling onto it. An optical filter is placed in front of each pick-up to correct for deficiencies in the dichroic mirrors.

Because each CCD is operating in the same manner as a monochrome image device, the resolution and light level performance of a three-chip colour camera are comparable to its monochrome counterpart. However, the combined cost of the optics and the three CCD chips makes this type of

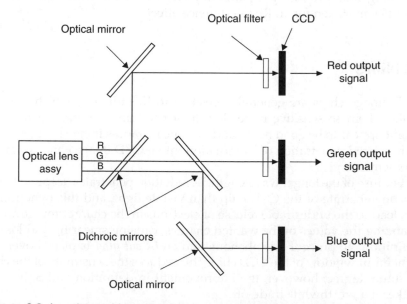

Figure 6.8 *In a three-chip colour camera, the light is broken down into its component parts and focused onto three CCD chips*

camera very expensive and has for many years assigned these cameras largely to broadcast and semi-professional video production use. As advances in technology bring the cost of three-chip cameras down we have, in more recent years, begun to see some three-chip cameras employed in CCTV applications; however, they are still much more expensive than their single-chip counterparts.

Colour cameras for the CCTV industry are generally single-chip units. The chip is identical to that used in a monochrome camera; however, a filter is placed in front of the CCD window to break the light up into red, green and blue. There are two types of filter in use: the *striped filter* and the *mosaic filter*. Each of these has its strengths and weaknesses, and comparing the two, the striped filter offers better colour reproduction, whereas the mosaic filter offers superior resolution.

The striped filter, illustrated in Figure 6.9, forms a mask of alternate red, green and blue strips of filter material in front of the CCDs. Each CCD produces an output for just one colour. These output signals are processed in the same manner as we saw for the monochrome chip in Figure 6.3; however, the output from the horizontal shift register has to be further processed to derive the luminance and colour signals.

The reason for the poor resolution with this type of pick-up is apparent when we look at Figure 6.9. In the horizontal direction, three CCD cells are required to produce one pixel, whereas in a monochrome pick-up the same cells would produce three pixels. It might therefore appear as though the

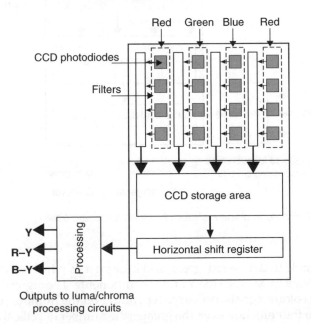

Figure 6.9 *Frame interline transfer (FIT) chip with a striped colour filter*

horizontal resolution is reduced by two thirds; however, when other factors are taken into consideration it can be shown that the resolution of this type of chip is about 50% of that of a monochrome chip.

Another problem which occurs with the striped filter is patterning on areas of fine picture detail. This is caused by the generation of unwanted frequency components when the picture detail falling onto the chip is about the same size as the filter stripes. This interaction between the filter and the picture information, known as *beating*, is overcome by placing a crystal filter between the main optical lens and the CCD chip.

The mosaic filter is considerably more complex, both in terms of the filter construction and the signal processing required. A filter is placed above each CCD cell in a mosaic pattern as shown in Figure 6.10. This sixteen-part block pattern is repeated across the entire chip.

Using the three secondary colours yellow, magenta and cyan, plus the primary colour green, the processing block following the horizontal shift

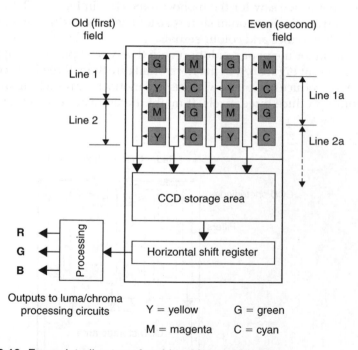

Figure 6.10 *Frame interline transfer chip with a mosaic filter*

register is able to derive red, green and blue signals which can then be further processed to produce the Y and C components. The algorithms used to derive the colour signals can vary, depending on the number of CCD cells there are in the chip; however, the greater the number of cells, the better the CCD performance.

Both the striped and mosaic filters reduce the light input to the CCD cells. This is one of the prime reasons why colour cameras are much less sensitive than monochrome cameras. While it is common to find monochrome cameras with a specified minimum light input level of 0.1 lux, in general the minimum level for colour cameras is in the order of 1 lux.

Camera operation

The principle of the colour camera is illustrated in Figure 6.11. The shift registers which process the CCD charges are contained within the CCD chip, and thus the input to the signal processing block is the output from the CCD horizontal register. The charge transfer process is controlled by the CCD

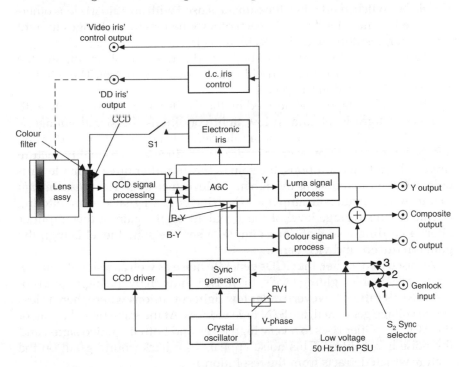

Figure 6.11 *Block diagram of a single chip colour camera (PAL)*

driver, which in turn is clocked by an accurate crystal oscillator. Because of the relationship between the charge transfer and the line and field sync signals, the CCD driver takes a reference from the sync pulse generator.

The signal processing is highly complex, but as far as we are concerned it is sufficient to note that there are three signals present at the output: the luma and the two colour difference signals R-Y and B-Y.

The average amplitude of the Y signal is determined by the average light input level, and thus the Y is monitored by the EI circuit. If the light input level is consistently high or low, the EI control changes the exposure time of the CCD accordingly. This function enables a camera to be fitted with a manual iris lens and yet maintain a degree of control under changing lighting conditions. However, it should be remembered that, unlike the manual iris, EI does not affect the depth of field and so, should the engineer set the manual iris to a low F-number, the EI will 'close' to maintain a correct light level; however, the depth of field will be poor. The situation will become even worse at night when the depth of field further reduces despite the fact that the EI has 'opened'. Another undesirable effect of electronic iris is the increase in smear when the exposure time of the CCD is reduced.

The EI circuit can be switched on or off by S_1, which may be accessible either on the side of the camera or through a software set-up menu. The EI should be switched off when the camera is used with an auto iris lens otherwise the two may tend to fight each other when sudden changes in light level occur, resulting in an iris oscillation effect.

The luma signal is also fed to both the d.c. iris control circuit, and the output socket on the camera provided for 'video' iris control. The action of these was discussed in Chapter 4.

The signal voltage level derived by the CCDs varies considerably with changes in light level, and in low light conditions the signal voltage is so small that a large amount of amplification is necessary to produce an acceptable output. However, in order to produce a high-quality picture under all lighting conditions, the gain of the amplifier must be made variable so that it can reduce as the light input level increases. The *automatic gain control* (AGC) circuit is an amplifier which includes a signal level monitoring circuit. As the average level of the signal alters, the gain of the amplifier alters accordingly so that, for example, when the signal level is high the gain is reduced, and vice-versa.

As discussed earlier, the CCDs generate noise as well as a signal voltage. Under reasonable lighting levels this noise is hidden by the high signal-to-noise (S/N) ratio. However, under low light conditions where there is less signal voltage generated, the S/N ratio reduces. At the same time the gain of the AGC amplifier rises to a very high level, and both the video signal and the noise is amplified. This noise appears as a background grain on the picture which detracts from the resolution.

As we have seen, the luma signal is derived in the CCD signal processing circuits; however, further processing is required before the luma is ready to be sent to the monitor. Much of this processing is of little interest to the CCTV engineer; however, one process which is worth mention is *gamma correction* as it has a direct bearing on the shape of the output waveshape.

CCD pick-up devices generally have a linear output response, and so equal increments of brightness give rise to equal changes in output voltage. Unfortunately the CRT characteristic is not linear but follows a curve like that shown in Figure 6.12. If we were to apply an equal increment staircase waveform (like that illustrated in Figure 5.9, Chapter 5) to the cathode of a

CRT, the light output would not rise in equal increments. Instead, the steps in the dark and light regions would give rise to large increments in CRT beam current, whereas the intermediate (mid-grey) steps would produce compressed levels with little contrast difference between bars.

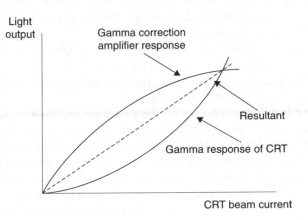

Figure 6.12 *CRT response and associated gamma correction curve*

Because of this the luma is passed through an amplifier whose characteristic is equal but opposite to the characteristic of the CRT. This process is called *gamma correction*. The effect of gamma correction on the luma staircase waveshape is shown in Figure 6.13.

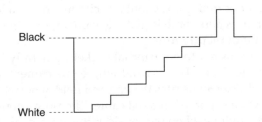

Figure 6.13 *Staircase signal with gamma correction (exaggerated)*

The gamma correction circuit could have been built into the monitors, but when we consider this from the point of view of broadcast TV where there may be just one or two cameras whose output is being viewed on literally millions of receivers, then it does not make economical sense to put a circuit into every TV. Therefore, the correction circuits are found in all video cameras.

Switching the gamma on and off on a CCTV camera appears to simply alter the contrast level; however, remember that with the gamma turned off the CRT will produce greater contrast changes at the lower and higher brightness levels, leaving the mid-grey levels somewhat bland. As a general rule the gamma correction should be turned on, offering a linear contrast range. However, you may experience cases where the CRT phosphor characteristic does not match that of the gamma correction in the camera and the contrast range appears too stark at the high and low levels. In such cases a more linear appearance may be obtained with the gamma correction turned off. The point to remember is, don't simply use the gamma correction setting as an alternative gain control.

Referring once again to the block diagram in Figure 6.11, the minimum bandwidth of the luma process circuits must be 0–5.5 MHz for a resolution of around 550 TVL, although with modern technology many CCTV cameras are able to exceed this resolution figure.

Once the luma signal has been derived, line and field sync signals are added. These pulses are generated in the sync generator stage, which derives an accurate signal by dividing down the crystal oscillator output.

The need for synchronization between the camera and monitor was discussed in Chapter 5, but at that point we only looked at a single camera and monitor. Problems arise as soon as we try to send two or more camera outputs to a single monitor through a switcher or multiplexer because, unless steps are taken to lock the cameras together in terms of synchronization, at any instant in time the scan position of the two cameras would be totally different. Each time the switcher toggled, the monitor would have lock to the new scan position, the picture rolling and pulling as it did so.

Switch S_2 is a three-position switch with a common output terminal connected to the sync generator. Position 2 is not connected to any circuit, and when the switch is in this position the sync generator circuit simply 'free wheels' and generates an independent sync signal. This position is used where the switcher or multiplexer contains circuits that can lock the camera signals together, or where there is just one camera, or where the customer doesn't mind a lot of picture roll!

Position 1 is connected to an external socket (generally BNC), which is usually labelled 'genlock'. This is used where one camera, or perhaps the switcher or multiplexer, generates the line and field sync, which is then sent to each camera via a separate co-axial cable. The poor economics of using this in CCTV are obvious when we consider that each camera requires two co-axial feeds. Genlocking is normally only employed in broadcast television where multicore cables are used, and cable runs are relatively short.

Position 3 is only available on cameras that are either 230 Va.c. fed, or 24 Va.c. fed. A sample of the 50 Hz mains frequency is fed to the sync generator which locks to the zero transit point in each a.c. cycle. If each of the cameras is connected to the same phase of the mains supply, then they will be synchronized. This method of synchronization is called *line lock* (LL). Problems can occur on larger installations where cameras in different locations are fed from different phases of the mains supply; however, this can be overcome by adjusting the V-phase control RV1, which is able to shift the sync output from the generator circuit through 120°, which is the difference between any two mains supply phases.

Multiple camera synchronization will be looked at in more detail in Chapter 9.

The colour signal process block takes the R-Y and B-Y signals and applies weighting to produce the u and v components. It also produces the 4.43 MHz sinusoidal subcarrier which, in the case of Figure 6.11, is done by dividing down the master crystal oscillator, thus ensuring that the phase relationship between the chroma, sync and CCD output is always correct.

The u and v signals are amplitude-modulated onto the subcarrier. Normally with AM only one signal can be put onto a carrier, but in a PAL colour TV signal a method called quadrature amplitude modulation (QAM) is employed which enables two separate signals to be placed onto one carrier in such a way that they can be separated in the colour decoder.

To complete the PAL chroma signal package, a sample containing ten cycles of 4.43 MHz subcarrier is positioned to coincide with the back porch period of the line sync pulse. This signal, known as the *chroma* or *colour burst*, is used to synchronize the colour decoders in the monitors or recording equipment.

The majority of colour cameras offer a choice of either separate Y/C outputs (S-VHS) or composite video. The relative merits of these were discussed in Chapter 5. All output connectors are terminated at 75 Ω to provide correct impedance matching to the 75 Ω co-axial cable.

White balance

Up to this point we have assumed that the CCD chip or chips in a colour camera are able to produce exacting proportions of R, G and B for any given light input. Unfortunately this is not the case. Tolerances between chips, and within each individual chip, affect the R, G and B levels to such an extent that the colour quality of the reproduced picture would be at best poor, and in most cases totally incorrect.

Following power-up, the camera needs to be shown what white light looks like. It then uses this information to derive correction values for the processing circuits which ensure that the correct proportions of R, G and B are produced for a white light input. In theory, if the camera is producing the correct proportions of R, G and B for white, then all other colours will be correct.

To perform white balance set-up a white card is held in front of the lens to provide a white light input. The white balance function is then selected, and the camera quickly adjusts its circuits and memorizes the correction factors. As an alternative to using white card, if there is a large white object within the field of view, the camera can be trained onto this.

This method is not always practical for CCTV applications and therefore the majority of modern cameras have an *automatic white balance* (AWB) function which uses pre-set correction factors. There is normally a choice of settings such as 'Indoor', 'Outdoor', 'Fluorescent' or 'Auto'. The manufacturer takes the average lighting conditions for these situations and designs suitable correction factors into the camera. Apart from operating the selector switch, there is no setting up required with this type of white balance, but there are obvious limitations with this technology.

The main drawback with automatic white balance is the effect that the different light colour temperatures have on picture quality. Take, for example, an external camera which is to rely on halogen lighting during the evening. If it is balanced using daylight as the light source, the colour may well be incorrect during the evening. Even during the daytime, difficulties may be experienced if areas within the field of view have different colour temperature lighting. This situation can arise when fully functional cameras are employed and their panning action takes them across different lighting conditions. These problems associated with AWB are one reason why for many years colour cameras were often considered unsuitable for external applications.

Advances in technology have largely overcome the problems associated with AWB. An automatic tracking circuit is included in many colour cameras, which ensures that once AWB set-up has been performed, correct white balance is maintained for all colour temperatures of light.

Back light compensation

A problem that is commonly encountered with CCTV cameras is where the user needs to record (clearly) a person when they are in a doorway. During the evening this is not too difficult; however, during the day when the area outside the doorway is brightly lit, the iris will close down to prevent overexposure of the outside area. Unfortunately this results in the person becoming underexposed and appearing as a silhouette. A similar effect was discussed in Chapter 4 (see Figure 4.12a); however, in this case the problem was resolved using the peak level adjustment in the lens.

Unfortunately many CCTV camera lenses do not have this facility and therefore most cameras incorporate a feature known as *back light compensation* (BLC). When turned on, the camera forces the lens iris to open up (assuming that an AI lens is fitted) to a point where the area outside appears overexposed; however, the person is at the correct exposure level and is therefore discernible. In practice this feature performs better on some cameras than on others, but for many years never really gave a satisfactory result. With newer technologies such as DPS and improved digital signal processing within the cameras, the problem of filming objects under these conditions is becoming much easier to resolve.

Colour/mono cameras

In order to ensure faithful colour reproduction in a colour camera it is imperative that no infrared light is permitted to enter the image chip and, to

Figure 6.14a *During the evening an acceptable image of the person is obtained*

Figure 6.14b *During the daytime the iris closes down in proportion to the light level around the doorway, resulting in the person having a silhouette appearance*

this end, all colour cameras have an IR filter in front of the image chip. Some monochrome cameras also have such a filter in order to prevent overexposure during the daytime caused by the large amount of IR present in natural sunlight.

The necessity of the IR filter in a colour camera did, for many years, limit their use to locations where the lighting conditions are relatively stable and where artificial lighting is both in the visible light spectrum and is of a reasonable intensity. This is why for many years only monochrome cameras could be employed where artificial IR lighting was employed. However, more recent developments have spawned a new generation of cameras that are able to function as both colour and monochrome, depending on the lighting conditions. These colour cameras are able to automatically switch to monochrome operation when the light level falls below a predetermined value. To enable such cameras to function with IR lighting, when they switch to monochrome operation, the IR filter is mechanically pulled away from the CCD chip.

Such cameras are able to take advantage of the much more desirable colour picture when the light level is sufficient, but can revert to the more effective monochrome operation when low light conditions prevail. Note that because in black and white mode the IR filter is removed and the CCD chip is functioning in monochrome mode, these cameras have much higher resolution and sensitivity figures for monochrome operation than they do for colour.

Early versions of these cameras tended to be somewhat unstable under some conditions – for example, where car headlights intermittently illuminate the viewing area – but improvements in performance during recent years have made these cameras a popular choice for external operation.

Camera sensitivity

There are a number of ways in which the sensitivity of a camera can be measured and quoted, which is not helpful to the engineer when trying to decide between cameras of different manufacture, because the figures may not be comparable.

A true sensitivity figure is the maximum F-stop which produces a 1 Vpp video output when the camera is trained onto a specific test card under specified lighting conditions. The *AGC must be switched off*, and the manual iris is progressively closed down to the point where the video signal begins to fall from 1 Vpp. At this point the F-number is read, and quoted as the sensitivity.

However, this is not the figure that is often quoted; rather, the *minimum illumination figure* is used. This is a measure of the lowest level of light falling onto an object from which a 'recognizable' video signal can be produced. Typically these specifications appear in a catalogue something like, '1 Lux – 80% – F1.2'. This implies a light level at the object of 1 lux, where the object has a reflectivity of 80%, and an F-stop of 1.2 was used.

It is the definition of 'recognizable' which causes some difficulty, because it is left to each manufacturer's judgement. This is by no means implying that all manufacturers are attempting to pull the wool over installers and specifiers, but it does offer a window of opportunity to some who wish to market what in truth are low spec cameras, whilst quoting high sensitivity figures.

An important consideration when looking at minimum illumination figures is to see if the measurement was taken with the AGC turned on or off. With the AGC turned off the figure is going to be more realistic because the camera will not have been working 'flat out' to produce the recognizable video signal. Be careful when comparing one camera which has been specified with the AGC off, with another with the AGC on. This is not a true comparison because it is almost certain that the model with the AGC 'on' will show a much better figure that that with the AGC 'off', yet it may well be that the camera with the AGC 'off' is the better of the two. Unfortunately not all specifications state whether the AGC was on or off when the sensitivity was measured.

Camera resolution

The theory behind the resolution of a video image has been dealt with in Chapter 5 where we saw that ideally a 625-line system should be able to reproduce 780 horizontal pixels (Figure 5.4), but that in practice, because of limitations in technology, we opted for something less during the design of the original broadcast system. For a 625-line television system with 585 active lines, the maximum number of horizontal pixels along one line is 780 × 0.7(Kell factor) = 546.

The resolution of CCTV cameras is quoted in *TVL* (television lines). This figure refers to *the number of horizontal pixels a camera can produce along a distance equal to the height of the screen*. This point is illustrated in Figure 6.15.

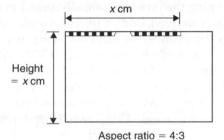

Aspect ratio = 4:3

Figure 6.15 *TVL is measured as the number of horizontal pixels along a distance which is equal to the screen height*

In Chapter 5 it was stated that, ideally, the horizontal resolution should at least equal the vertical resolution. Therefore if the maximum number of pixels along a line is 546 (PAL), then the number of pixels along a distance equal to the height will be 546 divided by the aspect ratio, which is 546 ÷ (4/3) = 409.5. Thus for a PAL CCTV camera to produce a picture which matches that seen on a broadcast transmission, it must have a resolution of at least 410 TVL. Similarly, for NTSC the horizontal resolution in terms of TVL will be 462 ÷ (4/3) = 346, although the practical figure quoted is 330 TVL.

In practice other losses in the system often require the CCTV camera to have a much higher resolution than the figures quoted above if the system is to match broadcast performance.

It should also be noted that whatever resolution is quoted, it will not apply to the entire picture area. Neither the lens nor the monitor CRT produce full resolution around the edges of the picture, and when looking at TVL figures it is prudent to assume that this performance will only apply to the centre portion of the displayed picture. This is of particular importance when selecting equipment for a specific operational requirement where the TVL for each image at the control room end has been specified.

The subject of camera resolution is considered in further detail in Chapter 13, when we look at methods of measuring system resolution.

Camera operating voltages

The three common supply voltages for CCTV cameras are 230 Va.c., 24 Va.c. and 12 Vd.c. Although at first glance the operating voltage may not appear too significant, the choice of camera supply has a direct bearing on the installation material costs and labour because some supply methods require more wiring than others.

For 12 Vd.c. cameras the supply is separately fed, meaning that every co-axial cable must be buddied with a d.c. supply cable. In some cases this is not a problem; however, where there is a need for all cables to be hidden, losing the extra one can sometimes prove difficult.

Another problem associated with 12 Vd.c. systems is that of voltage drop on cable runs exceeding (typically) 100 m. This phenomenon, along with methods of overcoming the problem, was discussed in Chapter 2.

The 12 V supply is derived from a power supply, the rating of which must be suited to the number of cameras being installed. The current drawn by a camera is typically between 350 and 500 mA and to avoid over-running a power supply, a useful rule of thumb is to look at two cameras per 1 A supply. From this it becomes obvious that for even a modest system a large power supply is required. There are two schools of thought on this subject. On the one hand a single 12 V power supply rated high enough to power all cameras can be employed. On the other hand a number of smaller units can be used, having the advantage that if one unit fails, the whole system is not put out of action. In practice it is not much more expensive to take the second option, and the problem of voltage drop can be reduced by dispersing the power supplies around the site.

230 Va.c. systems do not suffer the problems of voltage drop and the need for large power supplies. This supply is particularly suited to external camera applications, and is essential if a pan/tilt unit is employed as 230 V is required to operate the motor.

Another advantage of using the 50 Hz mains as a power source is that the line lock feature on the cameras can be employed to provide synchronization (see Figure 6.11).

The main drawback of the 230 V a.c. supply is the requirement of a fused spur at each camera location which can often make the installation more involved, especially where cameras are mounted on towers and must be supplied using an underground steel wire armoured cable. In the UK, current regulations require this work to be carried out by a competent person and final inspection and testing of the circuit must be performed, with a certificate of compliance issued to the customer by the inspector, in accordance with BS 7671. In other words, if the installer is not a qualified electrician, the electrical installation work will need to be subcontracted out, or the installer will run the risk of prosecution in the event of any mishap.

Another problem encountered with mains powered systems is that of ground loops, although this phenomenon is usually rectified using ground loop correctors. This was discussed in Chapter 2.

For internal use, 24 Va.c. cameras are becoming increasingly popular. Being defined as extra low voltage (ELV), 24 Va.c. does not come under the same regulations as 230 Va.c., yet overcomes the problems of voltage drop. As with 12 Vd.c. a separate a.c. power supply unit is required; however, some switchers incorporate a limited a.c. supply which is sufficient to operate a small system.

Some cameras are compatible for both 12 Vd.c. and 24 Va.c. operation. These cameras generally operate at around 9 Vd.c. internally and use an internal d.c. to d.c. converter power supply to reduce the 12 Vd.c. input, and an internal bridge rectifier and smoothing circuit rectify the 24 Va.c. input.

A variation of the low-voltage d.c./a.c. camera is the *line fed* system where the power to each camera is fed down the co-axial signal cable. The cameras are supplied from a 24 Va.c. supply contained within a dedicated switcher/controller. The video signal is superimposed onto the a.c. supply and the two are separated using filter networks within the controller. This system makes installation very simple, and even lends itself to the DIY market. However, the need for dedicated controllers and cameras means that there is little flexibility in system design, and the system cannot be extended beyond its maximum camera capacity.

Specialized cameras

In addition to the vast choice of monochrome and colour cameras available, advances in both CCD and digital technology have spawned a range of high-performance specialized cameras, particularly in low light and covert designs.

As we have seen, comparing like for like, monochrome cameras always outperform their colour counterparts in terms of low light ability and resolution, and high-performance designs are available with minimum illumination levels well below 0.1 lux. Thus the installer looking for a camera which will perform well in an environment with minimum illumination has never had such an easy task. Even with colour cameras the situation is changing rapidly. For many years it has been generally accepted that colour cameras are unsuited to external use because of their poor low-light performance, and the fact that they would only produce correct colours if the area was illuminated in such a way that it resembled a TV production set! Yet new designs in CCD image chips, along with the production of dedicated digital processing ICs, have enabled surveyors to re-think.

Nevertheless there are still circumstances where a quality image is required in situations where any form of artificial lighting, visible or otherwise, is unacceptable. In these circumstances a device known as an *image intensifier* may be employed.

Image intensifiers are not new, and have been used for many years to enhance the performance of cameras, including tube types which always struggled to operate under low light levels. They function by effectively amplifying the amount of incoming light using electronic means. The basic construction is shown in Figure 6.16.

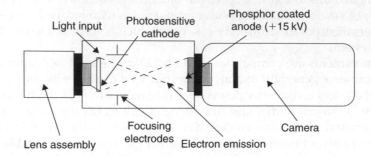

Figure 6.16 *Principle of the image intensifier*

The incoming light is focused onto a photosensitive cathode, the coating of which has the property of emitting electrons when impacted by photons. The electron emission from any point on the cathode is directly proportional to the amount of light falling onto that point, and thus there is a constant emission of electrons from all points on the cathode, representative of the image produced by the lens.

The electron cloud is attracted towards the anode by a high-voltage potential. As the cloud passes through the device it is focused electronically onto the anode plate, which is phosphor-coated in much the same way as a cathode ray tube in a monochrome monitor. Thus a picture of the original image is produced in front of the camera imaging device; however, it is many times brighter than that produced by the lens.

Use of an image intensifier makes it possible to obtain a clear picture with nothing more than starlight illumination. However, the cathode and anode materials have a limited operational life (typically 2000 hours), making image intensifiers a somewhat expensive option in any CCTV system. To obtain maximum lifespan, a high F-stop lens should be fitted.

Although image intensifiers are still in use, owing to their limited lifespan a more reliable alternative would be most welcome, and it is expected that developments in CCD technology will make them largely redundant.

Covert cameras

Covert cameras may be employed for any number of reasons, and not solely by the security forces. They are available for use by retailers, employers, entertainment centres and domestic homeowners alike. Some examples of use are: recording the activities of an intruder once they have defeated the overt CCTV system, production of video evidence which may be later used by either prosecution or defence counsels, installation in preference to overt equipment in order to maintain the aesthetics of the premises.

Until the advent of the CCD image chip covert cameras were relatively bulky and difficult to hide. Nevertheless, covert tube cameras were success-fully employed for many years. Of course it is not just the camera which needs to be compact; the lens must also be unobtrusive, and it is this which is perhaps the more difficult to achieve because the words 'miniaturizing' and 'optics' simply don't go well together. As we saw in Chapter 4, ideally a lens must be able to collect and process as much light as possible, and this means using large optical components. Having said this, small lenses offer-ing remarkable performance have been developed for covert use.

The most common type is the *pinhole* lens, although this name is some-what misleading because these tend to have a diameter in the order of 3.5–8 mm. They are generally available in ⅓" and ½" formats and have a CS mounting making them compatible with any matching format CS mount camera. Additionally, pinhole lenses are available in a right-angled con-struction to enable the camera to be hidden in locations where there is restricted space. This is illustrated in Figure 6.17.

Another type of lens developed for covert use is the mini lens which, as the name implies, is both short and narrow. However, the small size is achieved at the expense of certain components, and mini lenses generally have no iris, meaning that they are only suited to applications where the lighting conditions are reasonably stable. Note that a lot of mini lenses do not invert the image at the output, so the camera must be mounted upside down.

When selecting a pinhole or mini lens, the installer should try to choose one with the lowest possible F-number in order to attain the highest optical speed, bearing in mind that the optical speed will be restricted by the mini-aturized optics.

Fibre-optic lenses are available for use in situations where it is neces-sary to 'see' through thick walls or perhaps beyond a wide void. Both rigid and flexible types are available. The rigid type simply fixes to the front of a camera in the same manner as a conventional lens assembly, the light being carried from the small front optical device to the main lens assembly via a cluster of fibre-optic strands contained within a narrow tube. The flexible type allows the front optical plate to pass through more intricate spaces, and the camera does not have to be fixed adjacent to the field of view. Due to a number of types of losses within the fibre-optic strands, these lenses are not as sensitive as glass pinhole types.

Of course, all of the covert camera/lens arrangements so far discussed are expensive, which puts a lot of would-be users off. However, with recent developments in low-cost miniature covert combination camera/lens assemblies, the number of covert installations has risen considerably. Popular off-the-shelf covert CCTV products are the passive infrared (PIR) cameras (some of which include a working PIR), clocks, mirrors and domes-tic smoke detectors. 'Bare bones' PCB cameras are also available which can be mounted onto almost anything. Many of these are designed to be con-nected using standard four-core intruder alarm cable, which is far more flexible than co-axial cable and, despite the lack of screening and the

Figure 6.17 *Typical pinhole lenses for covert applications. A right-angled lens may be used where a covert camera is to be installed in a restricted space. (Photos courtesy of CBC (Europe) Ltd, manufacturers of Computar lenses)*

mismatch in impedance, the transmission results are usually quite acceptable, although most manufacturers do not recommend cable runs beyond 100 m. The camera modules are generally designed to operate at 12 V, which is delivered along two of the four cores.

360° cameras

Another technology that has become available as a result of modern processing power is a camera that has a 360° fish eye lens fitted, but is able to provide multiple corrected 4:3 images as though it were a number of separate cameras. The process includes a combination of high-resolution image sensors, complex mathematical algorithms and high-speed image processing. The detailed 360° panoramic digital image from the camera is 'de-warped' (straightened) and corrected to a 4:3 or multiscreen image, depending on user and application preferences. The images may be transmitted as a digital signal to a PC or NVR, or in analogue format as a standard PAL CCTV signal which can be monitored and recorded as normal. Figure 6.18 shows a typical camera head, along with the 360° panoramic and a quad display of four corrected images from that same panoramic.

Using a keyboard, the operator is able to view the selected areas, change the selected areas, pan around the 360° panoramic, or include the 360° panoramic within the screen multiplex. Alarm inputs are often available to cause preset areas to be called up. In some cases it is also possible to programme the unit to follow moving objects.

The example shown in Figure 6.18 shows a self-contained unit that includes all of the processing at the head end. An alternative to this is to

Figure 6.18a *Camera assembly with a 360° fish eye lens producing a panoramic field of view. An internal processor in the head assembly performs image correction and produces a number of multiplex output options*

Figure 6.18b *Panoramic view produced by the lens when the camera is ceiling mounted*

Figure 6.18c *Typical example of the video output available from this camera. Four corrected images in quad display created from the original panoramic image. It would be easy to believe that there are four separate cameras in the installation (Photos and images courtesy of Vista)*

employ a conventional CCTV camera with a 360° fish eye lens fitted and pass the panoramic image directly to a dedicated PC that is running a proprietary software. Image correction and multiplexing is performed by the PC, where the video information is also often recorded and stored. Although effective in terms of their ability to produce multiple, corrected images from a single panoramic, such solutions are not always as versatile as the head end solution because too often they are less able to integrate with other CCTV hardware.

The advantages of using these 360° cameras is that one unit can replace a multiple of static cameras, installation is greatly simplified, one single dome is aesthetically more acceptable than numerous static units, and the system can be made to record the 360° panoramic with the ability to view anything within that field at a later time. A typical application example is shown in Figure 6.19.

Number plate recognition cameras

Often referred to as vehicle number plate recognition (VNPR) or automotive number plate recognition (ANPR), these cameras are actually standard CCTV cameras. It is the software associated with them that performs the task of reading the number plate.

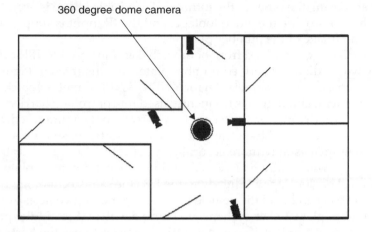

360 degree dome camera

Figure 6.19 *A single 360 degree, image corrected dome unit. This can replace multiple static cameras yet can achieve a greater range of image fields*

These cameras may be divided into two groups: those that use a standard visible light spectrum camera, and those that have an IR filter fitted in the lens.

Visible light versions simply provide a video feed to a PC that is running a software package which picks out the distinguishing features of a number plate on a vehicle. The software then displays (usually) a still shot and/or a short video clip of the vehicle, a close-up shot of the plate, and a computer-generated display of the plate. Provided that the camera is set up correctly, it is usually possible to identify the driver of the vehicle.

From the point of view of camera installation, for fixed installations the specifier or engineer only needs to consider the viewing angle and the required focal length of the lens, and (of course) the method of transmitting the video signal back to the monitoring centre.

In the UK, the fluorescent background on a vehicle number plate reflects infrared light very efficiently so, as daylight turns to dusk, the number plates may be illuminated using an infrared lamp mounted with the camera.

The one problem with visible light VNPR is that, because the camera has to look more or less directly at the vehicle, the vehicle headlamps usually blind the camera causing the iris to close, making the plate indistinguishable.

This problem may be overcome by using an IR filter lens. The camera is able to function in the daytime using natural IR light, and in the evening an IR lamp at the camera location switches on to maintain functionality, without the camera being affected by the headlamps because they are blocked by the filter.

The one major drawback with using IR is that the system can only provide a number plate. There is no useable video footage relating to the vehicle type or colour, or an image of the driver. To overcome this issue, VNPR systems normally take a feed from two cameras, one operating in the visible light spectrum, the other in the infrared. The visible image is used (where possible) to provide the video footage, and the IR input is simply used to derive an image of the number plate.

Typical images derived from both visible and infrared VNPR cameras, along with a digitally derived number plate, are illustrated in Figure 6.20.

The most common application for VNPR/ANPR is motor vehicle speed measurement, although it is frequently used for controlling barriers in private car parks. However, the technology is really only a simplified version of the more complex biometric video analysis software solutions that are becoming increasingly more accurate and effective, so it is clear that the technology will continue to evolve, and uses beyond vehicle plate recognition will be developed. Already in the UK, the same technology is being used to detect and read the vehicle tax discs on moving vehicles as well as the number plate. Other systems are being developed to aid the policing of the proposed policy of only permitting vehicles to use the 'fast lane' of a motorway if they are carrying two passengers – the cameras would be used to detect the presence of two persons (although this technology is, at the time of writing, somewhat controversial as it would, for example, be very difficult to prove the presence of a child on the rear seat). VNPR is also one method being looked at for the administration of toll road charges, although other 'vehicle tagging' methods are also being looked at in relation to this.

Figure 6.20a *Image from a VNPR camera with a visible light filter, taken during the daytime. The natural IR light provides sufficient illumination to derive an image of the number plate.*

Figure 6.20b *The same vehicle filmed with a colour camera (located next to the IR camera) fitted with a standard lens provides visible footage which may be used to identify the vehicle and (possibly) the driver.*

Figure 6.20c *Vehicle number plate, digitally derived from the images from the IR camera. (Photos courtesy of Vista)*

7 Video display equipment

In Chapter 6 we saw how the camera converts light energy into electronic signals relating to the luma and chroma content. In this chapter we will look at methods of converting these signals back into light energy, creating the illusion of moving coloured pictures.

For many years the sole device for producing coloured pictures for both the CCTV and domestic television industries was the *cathode ray tube*, or CRT. The vision to have a flat panel display has been around for many decades (in the 1960s *Star Trek* series, look at the futuristic flat panel 'main screen' that captain Kirk had on his bridge, a telling sign of where sci-fi buffs thought we would be in terms of video displays); it was the technology that was lacking.

Manufacturers, in particular Sharp Electronics, began working on LCD display devices a few decades ago; however, it was not really until the turn of the millennium that such display devices were becoming practical for displaying video images. There is a rapid shift towards flat panel (and in particular, LCD) displays because, we are told, they are 'better'. However, we must define the term 'better'. Compared with a CRT monitor, a flat display takes up less space, consumes less power, produces less heat, is lighter to transport, and looks modern. However, comparing like for like, a CRT monitor will provide much higher resolution and brightness levels for a fraction of the price. There is no doubt whatsoever that flat displays will equal if not outperform CRTs in the future. However (at the time of writing), where a high-resolution image is desired, a high-quality CRT monitor is still the best choice.

In this chapter we will look at each of the current display technologies and, where possible, compare their strengths and weaknesses. As it is the oldest form of TV display device, we shall begin with the CRT.

The cathode ray tube

People often refer to the thermionic valve as a thing of the past, something that went out during the 1970s with the rapid introduction of the transistor and silicon chip. Yet the cathode ray tube *is* a thermionic valve, as the diagram of a monochrome CRT in Figure 7.1 clearly shows. The operating principle is quite straightforward. The cathode, which is connected to a positive d.c. supply, is heated up. This causes the electrons present within its molecular structure to accelerate to an escape velocity, creating a negative space charge in the area just in front of the cathode.

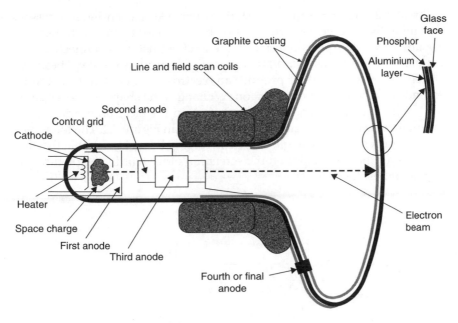

Figure 7.1 *A monochrome CRT*

A very high voltage, referred to as the extra-high tension (EHT), is connected via the final anode to the screen. The size of this voltage depends upon the screen size, but for a 30 (12″) monochrome tube, the EHT is typically 10 kV. If there were nothing between the cathode and the screen, the negative electrons in the space charge would make their way towards the EHT at the screen, and return to their power source via the final anode connection. Because electrons are moving from the cathode to the screen, we can say that there is a current flowing through the CRT. It is important to note that current flow only takes place because the tube has been completely evacuated of air. Otherwise the electrons would impact with the molecules of air and rapidly lose their momentum, causing them to fall back to the cathode once again.

The problem is that the electrons would not be travelling fast enough for our purposes, and furthermore, they would be spreading out as they travelled towards the screen due to the repelling effect of their like charges. To overcome these problems, anodes 1, 2, and 3 are added. The first anode (A1) is connected to a d.c. potential of around +100 V. This has the effect of attracting the electrons away from the cathode, giving them velocity in the direction of the screen. The electrons are travelling so fast by the time they reach this anode that they pass straight through a hole in its centre, and move at high velocity towards the EHT potential. Another name for the first anode is the accelerator anode. Unfortunately it can also be referred to as the screen grid. This is a name which goes back to the days of the valve, but today it can cause confusion for the unsuspecting engineer who might think that it makes reference to the screen on the front of the CRT.

Anodes 2 and 3 form an electrostatic focusing lens. Focusing is necessary to counter the divergence of the electron beam caused by the electrons moving apart. Anode A_2 is connected to the EHT, whilst A_3 is connected via a variable resistor (*focus control*) to a potential of around $+200$ V. These two largely differing potentials produce an electrostatic field between anodes A_2 and A_3. Altering the potential on A_3 changes the shape and strength of the electrostatic field, and hence the effect on the electron beam passing through the centre. The principle is illustrated in Figure 7.2. The focus control is adjusted until the angle of the converging beam is such that the electrons all hit the same point on the screen. In other words, the picture will appear at best focus.

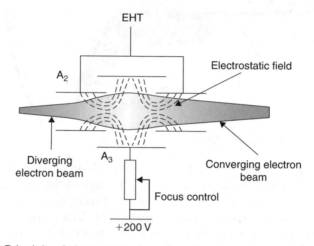

Figure 7.2 *Principle of electrostatic focusing. Cross-sectional view; the anodes are three metal cylinders*

The CRT screen is coated with phosphor, which glows when hit by the high-velocity electrons. Behind the phosphor is a thin aluminium layer, which serves both to protect the phosphor from burn, and act as a reflective screen to ensure that all of the light produced by the phosphor is projected forwards.

The assembly comprising the heater, cathode, control grid, and anodes is known as the *electron gun* assembly, although as we have seen, it does not actually fire electrons but rather it emits them as a result of the high EHT potential on the CRT viewing screen.

So we see how a CRT produces an electron beam current which flows from the cathode to the final anode, producing a small white spot in the centre of the screen. We shall now look at how the brightness of the spot can be controlled.

The brightness is dependent on the number of electrons hitting the phosphor, so by controlling the beam current we can control the brightness. This function is performed by the control grid, the action of which is illustrated in Figure 7.3.

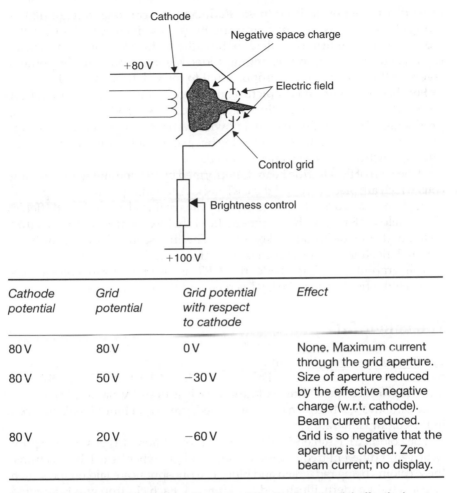

Cathode potential	Grid potential	Grid potential with respect to cathode	Effect
80 V	80 V	0 V	None. Maximum current through the grid aperture.
80 V	50 V	−30 V	Size of aperture reduced by the effective negative charge (w.r.t. cathode). Beam current reduced.
80 V	20 V	−60 V	Grid is so negative that the aperture is closed. Zero beam current; no display.

Figure 7.3 *Electric field produced by the grid/cathode potential effectively closes the aperture in the grid*

If the cathode is maintained at a constant potential, and the control grid is set at the same potential, there is no effect on the electron beam and maximum brightness still exists. However, making the grid less positive causes it to appear negative to the electrons leaving the cathode, resulting in those electrons close to the sides of the aperture being repelled. However, some electrons still pass through the centre of the aperture, and thus the beam current and brightness are reduced. When the grid potential becomes approximately 60 V lower than the cathode, the repelling action is so strong that the aperture in the grid is effectively closed. All beam current ceases, and there is no spot on the screen. Connecting a variable resistor to the control grid provides a means of brightness control.

The beam current can also be controlled by varying the cathode potential. Increasing the cathode voltage makes the grid appear more negative, causing a reduction in brightness. Reducing the cathode voltage makes the grid appear less negative, and the brightness increases. Thus, if the video signal waveform is applied to the cathode, the electron beam will be modulated by the rapidly changing video signal, producing brightness levels on the screen that are proportional to the video signal level.

For the CRT to produce a picture, the electron beam is made to deflect both vertically and horizontally at high speed, causing it to scan the screen, producing a blank white display known as a *raster*. Applying a video signal waveform to the cathode causes the beam current to be modulated, producing a picture as the screen is scanned.

Deflection of the electron beam is performed by the line and field scanning coils which are placed around the CRT neck. Alternating currents are passed through the scan coils which in turn set up alternating electromagnetic fields. These fields interact with a magnetic field that exists around the electron beam, and thus deflection takes place. The line scan coils perform horizontal deflection. The field scan coils perform vertical deflection.

The graphite coating inside the CRT is used to form connections between the final anode, the CRT screen and the second anode.

The colour CRT

In Chapter 3 we saw that white light is made up from the three primary colours: red, green and blue. Thus, for a colour CRT to work effectively it must reproduce these three colours, which it does by having three separate cathodes, each driven by separate red, green and blue signals derived in the colour camera.

The colour CRT screen is coated with three different types of phosphor which emit different frequencies (colours) of light when struck by electrons, these colours being red, green and blue. The phosphors are laid on the screen face in a tight pattern, illustrated in Figure 7.4. Each electron gun is targeted at just one set of phosphors, and so in effect we can say that one gun is creating the red light output, another the green, and another the blue.

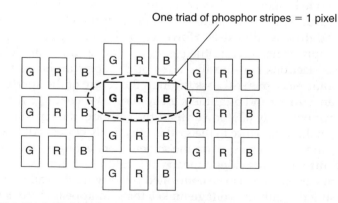

Figure 7.4 *Typical phosphor stripe formation on a colour CRT screen*

When all three guns are emitting electrons, all the phosphors are illuminated and the eye receives red, green and blue light. However, because the phosphor spots are so small, at normal viewing distance the eye is unable to discern them and the brain is tricked into thinking that it is seeing a white screen. The individual spots are visible if you move very close to a colour monitor screen and focus your eye on an area of white picture content.

Colour can be introduced by turning the guns on and off. For example, if the blue gun is turned off, the eye receives only red and green light and, as we saw in Chapter 3 (Figure 3.3), the brain will interpret this as yellow.

The principle of the colour CRT is shown in Figure 7.5. The actual operation is a little more complex than this diagram reveals because special magnets are required around the tube neck, either external or internal, to ensure that each electron beam only ever strikes its own colour phosphor. This is called *convergence* of the CRT, but it is not necessary to look at this in any more depth as it not something that the CCTV engineer will become concerned with.

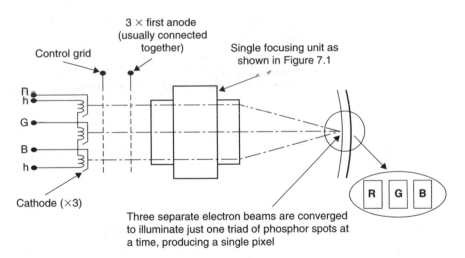

Figure 7.5 *Principle of a colour CRT*

The other main difference between colour and monochrome tubes is the operating voltage. For a colour CRT it is generally higher than for their equivalent size monochrome counterparts. The cathodes typically require a signal drive voltage of around 80 Vpp (swinging between 70 V and 150 V), the first anode potential is around 300–500 V, the focus is approximately 1 kV, and the EHT can be approximated as being equal to 1 kV per inch of screen size, and so for a 24" colour CRT the EHT will be around 24 kV.

CRT monitors

Generally speaking these have two video inputs: composite (CVBS) or Y/C, the latter being included to accommodate an S-VHS input. BNC connectors are normally used for the CVBS input/output, both because of their robustness and because they are designed to maintain the 75 Ω impedance of the transmission system, provided of course that the correct BNC type is used. BNC may be employed for the Y/C inputs; however, SCART or four-pin S-VHS connectors are commonly used.

Some monitors use RCA (phono) connectors for CVBS input/output, but this can present a problem when the monitor input cable is co-axial because RCA connectors, which were originally intended for audio use, do not fit onto co-axial cables such as RG-59. Thus a BNC to phono adapter must be used which creates an added connection in the transmission medium.

Owing to both manufacturing tolerances and viewing preferences, it may be necessary to perform a number of adjustments to a monitor during installation. Therefore we will look at the typical adjustments and their function. In practice most adjustments are performed in software using a menu; however, some are still performed the more traditional way using a terminal screwdriver.

There is often much confusion as to the difference between brightness and contrast. The brightness control moves the level of the entire image such that dark and light areas of the image all become lighter or darker. The contrast setting moves the light and dark areas towards or away from each other. The action of the two controls is illustrated in Figure 7.6.

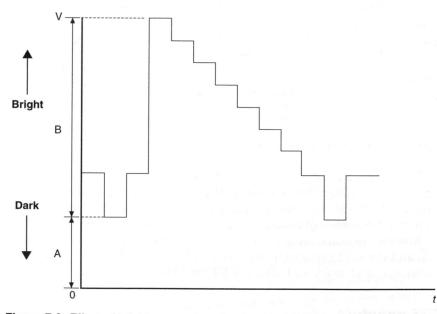

Figure 7.6 *Effect of brightness and contrast control. Brightness alters the d.c. level (A), moving the complete waveshape up or down without altering the amplitude. Contrast alters the peak–peak value (B)*

For example, turning the brightness level up fully results in the black areas becoming dark grey and the pale grey areas becoming white, with the white areas becoming excessively bright. Turning the contrast level up makes the dark grey areas become black and the light grey areas become white. Users will often adjust these to suit their own preferences; however, the service engineer should be aware that a CRT that is operating continually with the contrast set very high will suffer CRT failure very quickly (typically after two years). Also, in a CCTV control room the user may be complaining that the same images appear totally different when viewed on different monitors. This could be due to CRT ageing on some of the monitors, or the fact that they are of different manufacture, but it may be that the brightness and contrast settings at each monitor are set very differently.

In Chapter 5 we discussed the need for synchronization between the cameras and the monitors. Although modern manufacturing tolerances have removed the need for *vertical* and *horizontal* timebase adjustments, these controls are still found on some models. Incorrect adjustment of the vertical hold will result in field roll or picture bounce. If the horizontal hold is maladjusted, the picture may tend to pull to the right, or break up into a series of horizontal black lines. These effects are illustrated in Figure 7.7.

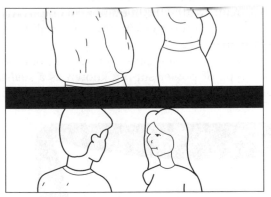

Vertical hold maladjusted

Horizontal hold maladjusted

Figure 7.7 *Effects of maladjusted timebase controls*

The *vertical* and *horizontal shift* controls are used to centre the picture on the monitor screen. The CRT *first anode* adjustment, also referred to as the *screen* or *A1*, is usually made available through the rear of the monitor. It is not recommended that this adjustment is performed by the CCTV engineer as the correct set-up procedure is somewhat involved. It requires removal of the cabinet, and a voltage measurement taken at the A1 terminal on the CRT whilst adjusting the control. The A1 appears to behave as a brightness control; however, it is not to be used for this function, and maladjustment of this can result in impaired picture quality, and a possible reduction in CRT life.

Adjustment of the *focus* control is also normally made through the rear of the monitor. The monitor should be made to display a picture containing a high degree of detail, especially in the centre, and the signal source should be known to be sound; in other words, the camera focus should be correct! The focus control is then adjusted for the sharpest picture at the centre of the screen.

Two other controls which may be available are *raster correction* adjustments. All monitors and TV receivers suffer from an effect known as *pincushion distortion*. This picture distortion, illustrated in Figure 7.8, appears at the edges of the display and is caused by differences in the distance between the point of origin of the electron beam (i.e., the CRT cathode) and points along the sides of the screen. Circuits within the line and field output stages are included to correct the effect, and on larger monitors especially, external adjustments are made available.

The monitor requires very stable d.c. supply rails, and for all colour monitors these are provided by a type of power supply known as a *switched mode power supply* (SMPS), which is employed because of its high efficiency.

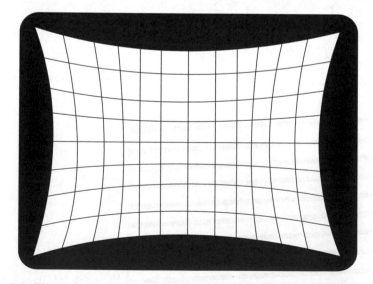

Figure 7.8 *Effect of pincushion distortion. Without raster correction, the edges of the picture become bowed*

The 230 Va.c. mains supply is full-wave rectified and smoothed to produce 320 Vd.c. This d.c. is then switched at a high frequency (40–80 kHz) through a transformer primary winding. Voltages induced in the secondary windings are rectified and smoothed to produce a variety of high- and low-voltage d.c. supplies.

The principle of the SMPS is shown in Figure 7.9, which illustrates a very important feature of this type of power supply: *mains isolation*. It is essential that the 0 V side of the d.c. power supply in a CCTV system is isolated from the a.c. mains supply. In items of equipment such as cameras, controllers and multiplexers, this isolation is performed by the action of the mains transformer. However, in the monitor it is possible for the 0 V line to find a return path to the a.c. mains supply through the bridge rectifier. If this were to occur, the chances are that the earth connections to the CCTV system would cause fuses in the monitor to rupture, as well as the RCDs to trip. However, if an earth fault were to exist in the system, all metal parts of the CCTV system may become live, with possible fatal consequences.

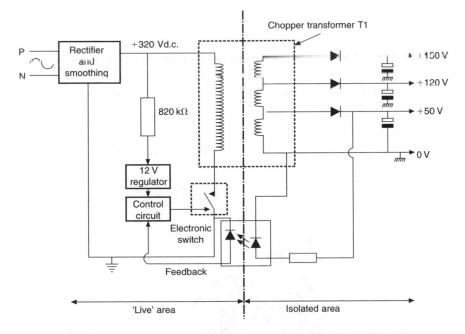

Figure 7.9 *Principle of the SMPs, illustrating the isolating effect of the chopper transformer*

By using the arrangement shown in Figure 7.9, the 0 Vd.c. line is completely isolated from the 'live' area by the transformer action. Even the feedback from the output to the control circuit, which is necessary to maintain a stable d.c. output, is isolated by means of an optocoupler. It is essential that no attempt is made to modify the power supply section in a monitor as this isolation may be compromised, and all servicing work in this section must be carried out by a qualified television service engineer, whose role and training is somewhat different from that of a CCTV engineer.

A word of warning: when a monitor is in stand-by, the SMPS is not switched off; it simply reduces the 150 V supply to such a level that the monitor is no longer able to function. This is done so that the +5 V supply to the CPU chip is maintained in order that it is able to receive and process a command to come out of stand-by. If an engineer were to remove the monitor covers whilst it is in the stand-by mode, there is the very real hazard of receiving a fatal shock from the 320 V rectified d.c., or the a.c. mains input.

Monitor safety

Monitor servicing should be referred to a qualified television service engineer, and the CCTV engineer should only perform the picture adjustments outlined in this chapter. However, there are occasions when removal of the monitor cabinet is necessary, and in such cases the engineer must be aware that this immediately exposes both him/herself and anyone in the area to the possibility of a lethal electric shock. Supplies such as the 150 V d.c. rail are able to deliver more than sufficient current to kill, and are especially lethal if applied across the chest, which can easily happen if the engineer has one hand on the chassis to hold it steady, and the other hand slips and comes into contact with a component connected to the high-voltage supply. This is illustrated in Figure 7.10.

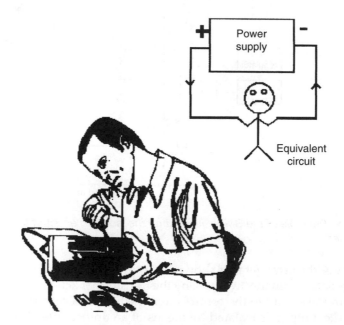

Figure 7.10 *Current flows from the positive supply (right hand), through the body and back to the negative supply (left hand)*

In this illustration the engineer has his left hand on the negative metal case (or chassis) of the equipment on which he is working. If his right hand were to slip and touch a high-voltage contact, a current path would exist between his hands, taking a path straight across his chest and therefore the heart. It has already been stated that this is a very dangerous situation.

Whenever working on a live monitor a 'one-handed' approach should be adhered to. This means that the engineer never has both hands on or in the monitor at the same time whilst it is connected to the mains supply, thus reducing the chance of electric shock across the chest. This point is illustrated in Figure 7.11, which shows the engineer working with one hand away from the chassis (his hand is on the bench; however, it could be placed on the plastic or wooden cabinet of the equipment). In this case there is no circuit and hence there can be no current flow.

Figure 7.11 *There is no circuit so current does not flow, provided that the bench top is insulated*

Of course there is still a chance of shock from the supply to earth, which brings us to the second important point on safety when working on a live monitor. *When working on live equipment connected to the mains supply, the equipment must be connected via a mains isolation transformer.* For a small colour monitor a rating of 500 VA is sufficient; however, for larger screen models a 750 VA transformer must be used.

It is not usually practical to work on a monitor on site. 230 Va.c. isolation transformers are not a usual item in a CCTV service engineer's kit, one-handed working can be difficult in what are often confined working conditions, and it can be difficult to ensure the safety of control room/area staff whilst the monitor is exposed. Consequently defective monitors are normally exchanged and serviced in a workshop.

In Europe, following servicing a monitor must be PAT-tested to prove its integrity.

One point which the CCTV engineer must always test is the integrity of the earthing on monitors which have a metal case. This includes not only the earth in the mains connector plug, but also the earth in the mains supply socket to which the monitor is to be connected. In a worse-case scenario a faulty monitor without an earth can cause not only its own case to become live, but the entire CCTV system including metal camera housings, mounting brackets, etc.

Liquid crystal displays (LCDs)

When it comes to an alternative to the smaller size CRT, the liquid crystal display is without doubt the most popular at the moment, mainly because it is much less expensive than its closest rival the plasma display (which we shall discuss later in this chapter). For the computer world the LCD is definitely a worthy contender to the CRT, but this is because computer graphics are not fast moving, which means that the refresh rate of the pixels does not have to be very high, and the user does not normally desire a very wide viewing angle. Once we require a viewing angle of around 200°, a refresh rate high enough to reproduce fast-moving objects without blur, and a high contrast ratio and brightness level, then only the more expensive and elaborate LCD panels begin to come close.

Early display devices employed a liquid crystal type known as dynamic scattering mode (DSM); however, these were not very efficient and were soon replaced by the *twisted nematic* (TN) type. TN crystal display devices are still the most popular, although other variants have evolved such as STN (super twisted nematic) and TSTN (triple STN). The basic TN device loses contrast as the number of scanning lines is increased, bearing in mind that PC monitors may produce many more scan lines than the 625 employed in television. STN and TSTN are capable of high contrast ranges at high line rates, even on larger display devices, although TSTN was really developed for large-screen monochrome display devices.

Liquid crystal is a substance that falls between a solid and a liquid. The long crystal molecules form up into a helix structure, and this has the effect that any light passing though the crystal experiences a polarity shift of 90°. When a voltage potential is placed across the crystal substance, the helix structure breaks down and incident light passes through unaffected.

LCD panels may be divided into two types: reflective and transmissive. Reflective panels use the control voltage applied to the crystals to determine the amount of light that is reflected off the crystals, whilst transmissive panels control the amount of light that passes through them. An example of a reflective display would be a pocket calculator where incident light is either reflected back off the crystals, or passes through them and is absorbed by the black back plate. TV display panels employ transmissive technology where light from a back light is either permitted to pass through the crystals, or is blocked by their action. One important point to note is that, unlike CRT or plasma displays, LCD does not produce any light and its operation is totally reliant on an external light source.

The principle of operation of an LCD panel is shown in Figure 7.12. If we ignore the crystal cells for a moment, light from the back light is subjected to the action of two polarizing glass plates. The vertical polarizing plate will only pass light waves that are vertically polarized, and the horizontal polarizing plate will only pass horizontally polarized light waves. Thus, the light output at the front screen will be zero because the first plate will block the horizontal light waves, and the second can only pass horizontal waves.

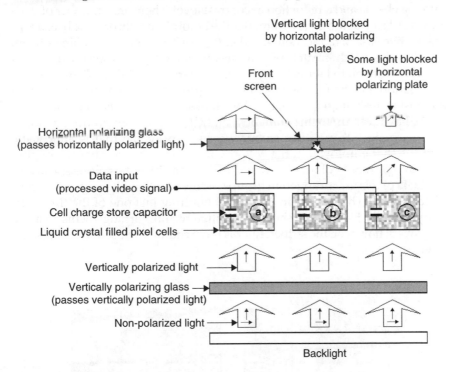

Figure 7.12 *Operating principle of an LCD display panel*

Via the data input, the liquid crystal cells are individually charged to a voltage potential that is relative to the video signal level for each pixel in the display. Consider pixel cell 'a' in Figure 7.12. When no charge is applied to the cell the helix structure will be intact and light passing through the cell is subjected to a 90° polarity shift, making it horizontally polarized. This means the light will be able to pass through the horizontal polarizing plate and a bright light output will be observed at that pixel location. At cell 'b' in Figure 7.12, a maximum charge is applied, the helix structure breaks down, the light passes through the cell unaffected, and it is therefore blocked by the horizontal polarizing plate. This equates to a black pixel location. For cell 'c', a 50% charge level causes a partial breakdown of the helix structure, meaning that the light is moved through about 45° and some of it will pass through the front plate. This represents a 50% light output at this pixel location.

In order for the cells to respond to rapid changes in brightness level, the switching devices that control the charge/discharge of the capacitors in the cells must be capable of very fast switching times. The devices used for such switching action are transistors that are fabricated on a thin film – hence their name, thin film transistors or *TFT*. Earlier LCD panels that did not employ TFT suffered a very slow response time, which resulted in blurring of fast-moving objects and a reduction in contrast level. There are a number of variations of TFT but the most common is the amorphous silicon type because of the relative ease of fabrication. Another type of TFT is polycrystalline silicon, which offers much greater refresh rates than amorphous silicon, although it is very difficult to fabricate, especially for large panel sizes. These devices have found a niche for use in LCD video projectors where high resolution and fast refresh are required, but the panel size need not be very large.

We have seen up to this point how an LCD display can produce a mono-chrome image. To produce colour we use a principle that is almost the same as that employed in the colour CRT. Remembering that we require the three primary colours (R, G and B) to reproduce the colour spectrum, the LCD pixels are arranged in groups of three, with a filter placed in front of each pixel. This results in each pixel emitting just one of the three primary colours. Figure 7.13 shows the different filter arrangements that are employed; however, the striped filter is most commonly used for TV display devices.

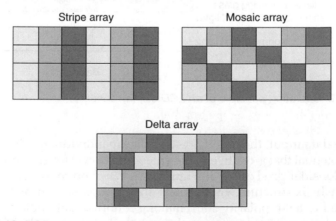

Figure 7.13 *RGB pixel cell array patterns employed in LCD panel displays. The stripe array is the most common type*

In Chapter 5 we saw that when an analogue video signal is converted into digital format, it is generally represented by 8-bit binary words, which gives us $2^8 = 256$ contrast levels. In an LCD panel, the cells are addressed using digital video signals, which means that the panels are (usually) capable of reproducing 256 grey-scale levels. For a colour display, bearing in mind that we have three different coloured pixels each capable of reproducing 256 levels, the maximum number of colours that we can obtain is $256 \times 256 \times 256 = 16.78$ million.

The number of pixels that a panel contains will vary depending on its size and the cost of the device. For example, an SVGA LCD panel would be required to have a pixel size of 800×600. However, bearing in mind that we require three LCD cells to make up one coloured pixel, the number of cells required would be $800 \times 600 \times 3 = 1.44$ million.

The back light is normally a cold fluorescent tube (although some larger display devices employ tungsten-halogen or metal-halide) that is driven by a high-frequency supply at around 500–700 Va.c. The drive frequency is carefully selected to ensure that it does not cause a strobe effect with the picture rate and is typically in the order of 40–60 kHz. The waveshape must be a pure sinewave because any harmonic components at such a high voltage could cause interference with the surrounding processing circuits, and even other equipment close by. Perhaps the most common failure of an LCD display is the back light because the tube itself has a limited life, and the power supply required to produce the drive voltage runs hot and as such, is more liable to failure.

Two problems that have plagued LCD display devices for many years have been the lack of contrast ratio and light output. In recent years manufacturers have devised some clever techniques to overcome these problems and, although each manufacturer has their own methods, the general principle is as follows. To increase the light output range, the display processor monitors the average light level for each picture frame. When it detects an overall bright picture, the processor increases the brightness of the back light. Conversely, for a dark picture content, the back light level is decreased. This technique can also improve the contrast ratio considerably.

The contrast may be further improved by employing a dynamic gamma control which also looks at the average brightness of each picture. When the average brightness is high, the controller increases the amplitude of the luminance signal and, conversely, for a dark picture the luminance level is reduced. By altering the luminance level we are in effect altering the contrast, but it is in relation to the average brightness level, so the changes are not apparent.

Another issue with LCD panels is the poor viewing angle when compared with CRT or plasma displays. For many years the viewing angle was very poor (and still is on some many less expensive panels); however, methods have been developed which enable a form of a lens to be fixed over each pixel, which disperses the light over a wider area. Thus, viewing angles in the order of 140° are now attainable.

Plasma display panels (PDPs)

Plasma display devices have been in use for a number of years and, in the domestic television market, have been challenging the LCD device. For a few years the playing field levelled out into one where LCDs cornered the smaller flat screen market (mainly because it was not possible to manufacture large size LCD displays economically) and plasma dominated the large flat screen market. In the early years plasma displays were very expensive and yet suffered from a lack of definition, poor contrast and a relatively short life (approximately three years).

In more recent years the field has changed; larger LCD displays are now available (with much improved picture quality) and, thanks to improved manufacturing techniques and image processing, plasma displays offer high resolution, improved contrast range and much longer life expectancy. At the time of writing, it remains to be seen where the market will go, and whether the LCD ultimately will manage to dominate the market as it currently threatens to do.

The principle of the plasma display can be seen when we consider a conventional fluorescent tube which, when subjected to a high electrical charge, glows because the gas inside breaks down into a plasma. Each pixel in a plasma display device is filled with a mixture of xenon and neon gas and has electrodes fixed to it so that, when a voltage is applied across the electrodes, the gas breaks down and some of the electrical energy is converted into electromagnetic waves, i.e., light. The light produced is in the UV spectrum; however, it is converted into visible red, green and blue by the addition of phosphors inside each cell. The phosphors are excited when hit by UV light and in turn emit visible light energy.

Each pixel is made up from three plasma discharge cells where each cell contains one type of phosphor. Therefore each pixel contains a red, a green and a blue light-emitting cell. The basic construction of a single pixel is shown in Figure 7.14, where it can be seen that each cell actually has three electrodes. We will consider the function of these in a moment. The phosphor linings can be seen in the diagram and, when the gas discharge takes place, the phosphors emit visible light out through the front viewing screen. The electrodes that are attached to the front glass are made from a transparent material so that they are not visible on the screen.

To produce different coloured pixels it is necessary to prevent some of the cells from firing. For example, if a certain pixel needs to appear yellow, then only the red and green cells must fire. The data electrodes control the firing of the cells by having their driver set to either a logic 1 or 0. Only those cells that have been addressed as 1 will fire during the discharge period.

The sustain electrode determines how long the cell will maintain its discharge. The longer the discharge period, the brighter that cell will appear. The scan electrode works in conjunction with the sustain electrode to produce the discharge potential and provide a discharge current path.

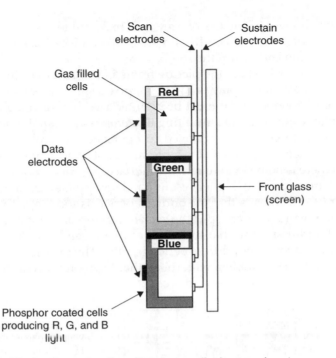

Figure 7.14 *Pixal cell construction of a plasma display panel*

For each TV frame, the display progresses through a three-step cycle: set, address and discharge (sometimes referred to as initialize, write and sustain).

Bearing in mind that the cells will still be illuminated from the last TV frame, at the start of the next frame it is necessary to ensure that all cells are extinguished and are restored to a neutral state. To accomplish this, a set pulse is applied between the scan and sustain electrodes, which causes the gas to produce a small, low-level discharge.

Following this short set period, the address period begins. This is when the digitized video signal is used to determine whether or not a cell should fire by applying a logic level to its electrode control circuit. For each cell that is set to logic 1, a negative pulse is applied to the data electrode and a positive pulse to the scan electrode. This has the effect of charging the gas, in which state it does not actually emit any light energy. The address period lasts for approximately 2 ms, during which time all cells will have been addressed and charged as necessary.

Following the address period comes the discharge (or sustain) period, when all of the cells that have been charged are simultaneously discharged to create the picture frame. Discharge is achieved by pulsing both the sustain and scan electrodes in such a way as to create a high potential across the gas, causing it to break down and emit light energy. The brightness of the cell is determined by the length of time that the sustain pulses are applied, which in turn is determined by the binary value of the digital video signal corresponding to that pixel cell. As with LCD panels, the norm is to use 8-bit video which gives 256 levels of grey and 16.78 million colours, although some monitors may employ 10-bit video, which gives 1024 levels of grey and 1.07 billion colours (in theory!).

Unlike the CRT where the picture frame is scanned over two consecutive fields, a plasma display produces all pixels simultaneously. Therefore the incoming video signal must be de-interlaced before being applied to the display control circuits, resulting in a *progressive scan*. Remember that interlaced scanning was employed to eradicate the problem of picture flicker associated with the CRT. A plasma display does not flicker in this way, so there is no need to retain the interlace.

Problems occur with higher-resolution display panels because the large number of cells means that the address period is too long. To overcome this, some manufacturers split the screen into top and bottom halves (in terms of addressing) so that two rows of pixels can be addressed simultaneously, reducing the address period by 50%. The principle is illustrated in Figure 7.15. The problem with this is that the monitor requires two display driver circuits, adding to the cost.

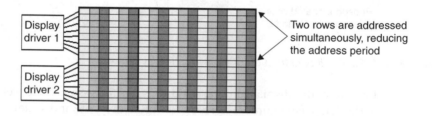

Display driver 1 — Display driver 2 — Two rows are addressed simultaneously, reducing the address period

Figure 7.15 *By dividing the panel display into two halves and employing separate display drivers, two pixel rows can be addressed simultaneously*

A problem that may be experienced with PDP devices is a poor contrast range, which is a result of their lack of ability to produce a true black level in the image. This problem is due to the low-level light emissions resulting from the necessary discharge of the cells (by the action of the scan pulse) at the end of each frame period. In more recent years some manufacturers of PDP devices have employed ingenious methods of modifying the scan pulse action such that the cell discharge is more gentle and therefore does not produce the same amount of UV energy.

Another problem that is inherent in PDP devices is an effect that is known as false contours. This effect only occurs on moving images, and usually only in brighter picture areas, where certain areas appear overexposed. The source of this effect can be traced to the coding used to set the discharge period in the cell drivers, where the light level content for adjacent cells becomes added together by the human eye as it follows a moving image across the screen. Once again, different manufacturers have found different solutions to this problem, some of which perform better than others, which is one reason why one PDP may appear to perform better than another. In general, the issue of false contours is overcome by controlling the discharge pulses in such a way that the eye cannot integrate the information in adjacent cells in bright moving objects.

Whether or not plasma display devices have any future in CCTV applications remains to be seen. Some control rooms have employed them as a means of providing large screens for general monitoring or multiplex display; however, this was, perhaps, before the availability of large-screen LCD. The future of PDP will be determined by its ability to develop and improve at the same rate as LCD and therefore remain competitive, both in terms of image quality and price.

Projection systems

Video projection is by no means a new art. Back in the 1950s black and white projection televisions were developed to provide a large-screen alternative to the 12″ television CRT which, at the time, was the largest available screen size. For many decades to come, projection monitors were large, heavy, expensive units that produced low-resolution coloured pictures, and offered a very narrow viewing angle and a poor contrast ratio. Hardly a contender for a CCTV control room! Advances in technology have moved projection monitors forward a long way, and in more recent years a number of models have become available that do offer a reasonable picture quality when viewed from a distance (which is how any large screen is meant to be viewed!).

If we compare any high-quality LCD or PDP display monitor with an equivalent-size projection monitor, the LCD/PDP models will always come out best in terms of contrast ratio, resolution and (usually) price. Where projection TV still has the dominance is in the very large screen market. Any CCTV control room that requires very large displays for monitoring purposes may find a solution in a projection monitor. However, both the engineer and the owner must be aware that the maintenance cost for these units tends to be higher than for LCD or plasma displays because of the lamp life. It is wise to check the specified lamp life for one of these units as, when lamp replacement becomes necessary, the replacement cost of the lamp unit alone will run into hundreds of pounds sterling at least, and can sometimes run into thousands of pounds sterling.

Projection monitors usually employ an LCD panel in the light engine to produce the red, green and blue pixels. The operating principle is quite similar to that of the LCD display, except that instead of having a relatively low output back light as a light source, projection monitors employ a high brightness xenon lamp with a light output in the order of 1000–2000 lumens.

An alternative to using a self-contained projection monitor is to employ a video projector. Here there are two competing technologies: the LCD and the *digital light processor* (DLP) projector, which is designed around a *digital mirror device* (DMD) light projection engine that is manufactured by Texas Instruments.

The basic DLP principle is illustrated in Figure 7.16, where the action of four mirrors is shown. Each mirror can be rotated through 45° by application of a drive voltage, and the drive voltage is controlled by a binary address which determines the brightness of the pixel produced by each mirror. With no deflection voltage applied, the mirrors are in their rest position and the light is reflected into a light dump area within the projector. When a voltage is applied to a mirror, the mirror tilts and the light is reflected out through the lens assembly. In Figure 7.16a, the four mirrors are shown in their rest position and thus there is no light output to the screen. In Figure 7.16b two mirrors have been rotated, and therefore their light output is projected onto the screen to produce two pixels.

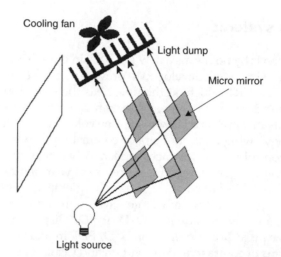

Figure 7.16a *DMD operating principle. With all micro mirrors set to their default rest position, all light is reflected onto the light dump*

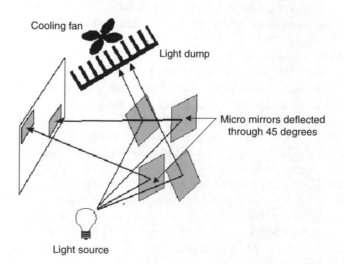

Figure 7.16b *Deflecting the micro mirrors through 45 degrees results in the light being reflected out through the front lens and onto the screen*

A practical DMD comprises at least 1.3 million tiny aluminium mirrors producing the pixel count required for a large-screen projected image. Grey scale is achieved by oscillating the mirrors during each TV frame period, the mark-to-space ratio of the oscillation determining the brightness of each pixel. For example, a 1:1 oscillation rate would produce a mid-grey level, whereas as 3:1 oscillation rate would produce a dark-grey pixel because 75% of the light landing onto the mirror would be reflected onto the light dump.

The light engine is the name given to the assembly containing the three DMD chips and the associated optical assembly centring around a large prism (Figure 7.17). The function of the prism is to split the white light into its red, green and blue components and focus the three outputs onto the corresponding DMDs. The light reflected from the DMD mirrors then passes back through the prism and into the main optical lens assembly or onto the light dump, depending on mirror orientation. The principle is shown in Figure 7.18.

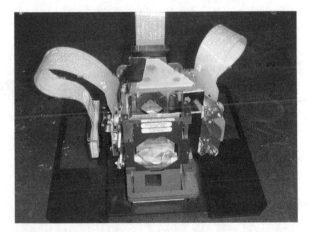

Figure 7.17 *The light engine comprises a light prism, three DMD devices, three DMD driver cards, three thermal electric coolers (TECs), the light dump, a mechanical shutter and a cooling fan*

The main advantage of DLP over LCD projection devices is shown in Figure 7.19, where it can be seen that the density of the mirrors is much higher than that of the LCD cells. This is due to the fact that there is no need for any separator between pixel devices in a DLP engine. This higher mirror density results in a very efficient light output. DLP projectors usually employ modified mirror drive modulation techniques similar to that used in plasma displays to provide increased contrast ratio and to counteract unwanted effects caused by the human eye integrating the light from adjacent pixels.

Although video projectors are capable of producing high-resolution, large-scale images, this is at some considerable expense. Furthermore, because

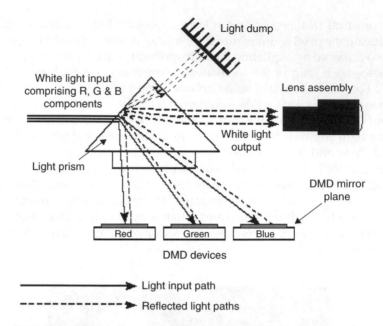

Figure 7.18 *The light prism separates the R, G and B light components and focuses them onto their corresponding DMD device. The reflected light paths then pass back through the prism where, depending on the angle of each micro mirror, they pass on to the main optical lens, or into the light dump*

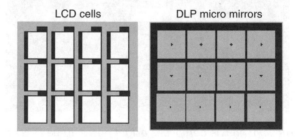

Figure 7.19 *DLP technology enables the mirrors to be packed more densely than LCD cells, resulting in greater light output efficiency*

of the noise and heat that they generate, plus the fact that they must have a clear, unobstructed view of the screen, video projectors are not a popular choice for normal viewing applications in a CCTV control room.

Termination switching

In Chapter 2, Figure 2.6, we saw that for maximum power transfer to take place between any two devices, the output impedance of the first device must equal the input impedance of the second device. When the impedances are not matched, some signal loss will be evident and, in the case of a signal transmission system, signal reflections may occur. Figure 7.20a shows a passive circuit arrangement for composite video input/output connection on any item of CCTV equipment. With this arrangement, the input and output sockets are effectively connected in parallel across a resistor, meaning that changes in impedance at the output will alter that at the input. Therefore the input impedance, which should be 75 Ω in order to match the co-axial cable, is dependent on whether or not another piece of equipment is connected to the video output socket.

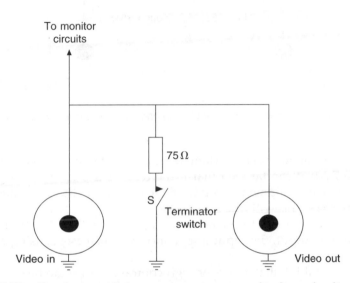

Figure 7.20a *Terminator switch arrangement on a monitor loop circuit*

In Figure 7.20a, when an item of equipment is connected to the output socket, correct impedance matching will be maintained by the input circuit of that equipment. When there is nothing connected to the output socket, a switched 75 Ω resistor is made available to maintain correct impedance matching.

For many years this selector switch, known commonly as the *termination switch*, was manually set by the installation engineer. The correct setup method was to set every switch in a chain of equipment to the 'out' or 'off' position except for the one on the equipment at the end of the chain. In more recent years, this switching action has been automated by making it a part of the output BNC socket. When there is nothing connected to the output socket, the switch is closed, connecting the resistor into circuit. The action of connecting a BNC plug into the socket opens the switch contacts.

An alternative, and more effective, method for automatic termination switching is shown in Figure 7.20b, where active buffer amplifiers are employed at both input and output sockets.

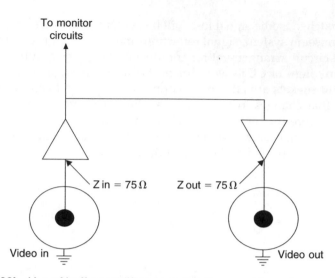

Figure 7.20b *Use of buffer amplifiers to maintain correct impedance matching*

Correct termination is important in a CCTV installation. Looking again at Figure 7.20a it can be seen that if the resistor is switched into circuit when an item of equipment is connected to the output, the impedance will fall below 75 Ω. This will cause an attenuation of the signal, resulting in a poor-contrast picture. If the switch is left open when there is nothing connected to the output, the input impedance will increase, causing a high-contrast display.

It is easy to think that a low- or high-contrast picture can be easily corrected by simply adjusting the contrast control on the monitor; however, there can be more serious consequences from an incorrectly set terminator switch. In the case where the resistor is switched in when it shouldn't be, the reduced signal is passed onto the next piece of equipment in the chain, and so on. Similarly, when the resistor is omitted when the monitor is the last item in the chain, the excess signal level may result in frame roll due to clipping of the sync pulses.

More serious consequences may become evident in systems where telemetry is passed down co-axial video cables (see Chapter 10). Remember that co-axial cables behave as LC tuned circuits, and that standing waves exist along their length. As we saw in Chapter 2, when the terminating impedance is incorrect, the standing waves alter, and reflections can pass back down the cable. Such a condition can cause corruption of the telemetry data, resulting in intermittent or permanent loss of control of remote cameras, etc.

Occasionally an engineer may encounter a situation where there is neither an input buffer nor a terminator switch. In such cases a 75 Ω terminator should be employed. These devices resemble a sealed BNC connector; however, they have an internal 75 Ω resistor.

Resolution

This is generally quoted in TVL, the definition of which was given in Chapter 6 (see Figure 6.15).

Monitor resolution is governed primarily by the quality of the display device, although the design of the drive circuits does have some bearing. A quick look through any CCTV supply catalogue reveals a range of monitors with resolutions from 300 TVL to over 1000 TVL, although as with cameras, high-resolution monochrome monitors are far less expensive than their medium-resolution colour counterparts.

When selecting a monitor for a particular application, resolution is one factor which must be taken into account. Yet it should be remembered that for CRT monitors, the TVL quoted will not apply to the entire screen area because a CRT suffers problems of beam de-focusing at the sides and, in particular, in the corners. In addition, colour tubes often suffer convergence problems in the corners of the screen. If we add to this the problems of image distortion around the edges introduced by the camera lens, then we see that it is highly unlikely that a monitor with a resolution of 400 TVL will be able to produce such resolution across 100% of its screen area. As a rule of thumb, if you imagine a rectangle in the centre of the screen with an area equal to about one third of the screen area, then this is where the quoted resolution will be achieved. Thus, if it is required to have a resolution of 400 TVL at the sides of the picture, then a monitor of around 450 TVL would most probably be required.

Ergonomics

Some CRT monitor designs are derived from domestic TV receiver models where the tuning and RF signal processing stages have been removed, and a range of video (and possibly audio) input/output sockets are added. Being designed around a domestic receiver, these monitors frequently offer high resolution. However, the moulded plastic cabinet does not make them particularly robust, and it is not possible for them to be stacked. The metal-cased rectangular design is often more suitable for CCTV purposes as it is more able to withstand the industrial environment, and it is possible to stack units into banks.

Careful thought should be given to the positioning of monitors during the initial site survey. Such things as viewing height, distance and angle, ventilation, monitor size, display type and resolution should all be considered at this point. Standards exist for the ergonomic design of control centres (ISO 11064: Ergonomic design of control centres), and these have considerable bearing on the design and layout of CCTV control rooms.

Generally speaking, monitors should be positioned so that they may be viewed without causing discomfort to the operator over a period of time. The distance of the monitor from the operator will depend on its size and function; however, recommendations are laid down in the new standards which are outlined in Table 7.1, along with the figures quoted in the BSIA Code Of Practice for CCTV (BSIA No. 109: Issue 2: October 1991). In addition to the distance, the viewing angle must not be excessive, and 30° maximum is recommended for monitors located away from the operator, and 15° maximum for a monitor mounted on a desk directly in front of the operator. Also, site the monitor so that the display is not impaired by glare from light sources, otherwise steps must be taken to remove or mask the source of the glare. Alternatively, an anti-glare screen can be fixed to the monitor, although this can reduce the light output.

Table 7.1 *Note that some of the BSIA figures relate to slightly different monitor sizes. Where this is the case, the monitor size is shown in parentheses*

Monitor size	Viewing distance (ISO 11064)	Viewing distance (BSIA COP)
23 cm (9″)	0.9–1.2 m	1–2 m
30 cm (12″)	1.2–3.0 m	2–3 m (39 cm/14″)
43 cm (17″)	1.3–3.6 m	–
53 cm (21″)	1.3–4.6 m	3–5 m (51 cm/20″)

One point to consider when stacking or racking monitors is ventilation. The reliability of a monitor may be reduced if it is made to operate continually at a high temperature, and in a large tightly packed stack the units in the centre could easily become overheated. Of course it is not always up to the installing or servicing engineer how the monitors are to be mounted, but where a monitor is found to be overheating, the hazards associated with this (unreliability, fire, etc.) should be pointed out to the customer.

The required monitor size is dependent on such factors as the amount of detail which is to be discerned and the number of images that are to be displayed. For example, it would not be practical to display sixteen images on a 23 cm monitor.

Finally, make sure that it is possible to access the monitor after installation. It will be necessary for operators to routinely clean the screen, as well as make occasional adjustments and, when a monitor fails, it will be necessary for the service engineer to remove it and install a temporary or permanent replacement.

8 Video recording equipment

The ability to be able to record CCTV images is of immense importance because it not only enhances the effectiveness of systems which are at times unmanned, but it also provides supporting evidence for control-room operators who by nature of their work become eyewitnesses to the incidents they are monitoring. On the one hand the video evidence will only be as good as the system performance will allow; on the other, if the specification of the recording equipment does not match that of the system, or the equipment is incorrectly set up or is poorly maintained, then the quality of evidence might well fall far below the live picture quality offered by the system.

It was not until the VHS (video home system) machine arrived on the scene during the latter part of the 1970s that the security industry had anything like a viable recording system to complement CCTV installations. Prior to that, the only recording medium available to the CCTV industry was reel-to-reel machines which only recorded in black and white, and were not particularly successful owing to their high cost, short recording time, low resolution and high maintenance requirements. Further modifications to VHS resulted in time-lapse recording and an enhanced VHS format – Super VHS – which provided the industry with equipment that offered acceptable recording duration and picture resolution. However, the issues of relatively high maintenance requirements and tape management remained, and, as CCTV systems increased in size, these issues became increasingly apparent.

Digital video recording has been available to the CCTV industry since the early 1990s; however, for many years the large file size of the recorded TV frames coupled with limitations in disk capacity and processor power meant that the recording time was very much limited and the frame update rate was poor.

Consequently these machines still relied heavily on DAT (digital audio tape) recorders to archive footage and, as DAT was only a variant of VCR technology (it actually evolved from Sony's early Betamax technology), many installers and users decided to stay with S-VHS. A compromise between analogue tape and hard disk recorders was found in the various digital tape formats that became available in the latter part of the 1990s. Two typical examples of these were the digital time-lapse (D-TL) format developed jointly by Panasonic and Sanyo, and Sony's digital video (DV) format. However, despite the outstanding image quality and the fact that the quality did not deteriorate over time, there was still the issue of tape management to address.

By around the year 2000, computer technology had progressed to a point where more viable disk-based digital video recorders were available, and since that time there has been a rapid shift from analogue and digital tape to disk-based DVRs (digital video recorders) and NVRs (network video recorders).

In this chapter we shall examine the principles of video recording, concentrating on disk-based recording. However, because at the time of writing VHS is still not completely dead in the industry, at the end of the chapter we shall overview time-lapse VHS recorders.

Digital video recorders (DVRs)

A DVR is basically a PC that has been designed to perform one specific function, which is to take both analogue and digital video input signals, digitize the analogue signals, compress all signals as necessary, and record these signals onto a hard disk. The DVR enclosure may be of a custom design, or it may be a standard PC tower, but either way it still utilizes state of the art PC technology including a high-speed (often dual-core) processor, a large-capacity hard disk, a large amount of RAM, and an operating system such as Microsoft Windows or Linux. Because all video images in the DVR are in the digital realm, most machines incorporate a multiplexer and therefore, in reality, they are two units in one. Multiplexing principles will be discussed in detail in Chapter 9.

It is interesting to track the evolution of the DVR from its early beginnings in the early 1990s when, because of the limited PC technology of the time, it was very limited in terms of recording capacity, frame refresh rate and picture quality. As processor speeds and memory capacity continued to increase, hard disk size and reliability improved, and compression techniques evolved, then a steady improvement in DVR performance was observed. And there is no doubt that this trend will continue for the foreseeable future as PC technology continues to break new boundaries, and compression techniques continue to improve.

It is common to find manufacturers equating digital image quality with analogue recording performance – for example, stating that a file size of 20 kB will produce an equivalent picture performance to that of a S-VHS machine. Whilst it is understandable that they are attempting to use a familiar image quality to help the installer and end user appreciate the sort of quality they should expect for a given file size, in truth it is difficult to accurately equate analogue and digital images in this way, especially with the compression method used in MPEG. This is because the losses in digital recording appear somewhat different on the screen from those of analogue recordings.

Where MPEG video compression is employed, as the amount of compression is increased, the image begins to break up into pixel blocks, a feature that will never appear on an analogue recorder. Yet these blocks are often only evident in areas of very fine picture detail which would never have been recorded by an analogue machine anyway. Thus a comparison of the performance of the two formats may reveal that, although neither format is capable of reproducing a particular part of an image, the loss is manifest in quite different ways. In most cases a comparison of the areas of picture that can be resolved reveals that the digital images are cleaner and have a greater contrast ratio. On the other hand, digital images are proving more difficult to enhance using current techniques than their analogue counterparts, which is not good news for anyone involved with forensic analysis of video information.

So what makes one DVR better than another? Well, processor speed and type, RAM capacity and hard disk capacity all make a difference. The type of operating system can also affect the speed of the machine. But perhaps the most important consideration is the quality of the hard disk that is to be used to record the digital video information. One of the biggest problems with DVRs is hard disk failure because the disk is working flat out, 24/7. The larger manufacturers making custom DVRs will only use high-quality, server-class, hard disk drives that are designed for continuous operation. However, be careful when choosing to use DVRs that are based on standard PCs which have been adapted by the installation of a video capture card and suitable application software. Such machines must incorporate a high-quality hard disk drive if they are to function reliably for any length of time.

The user interface for a DVR varies considerably. A number of manufacturers have chosen to mimic the buttons on a VHS machine, which proves to be a very friendly interface for users who are not at all technical and may struggle to cope using a mouse and software application, whereas everyone knows how to use the basic functions on a VCR.

Where the DVR is based on an adapted PC, the user interface tends to be via the PC display. Some of these are more friendly than others, and should be tested before a final decision is made to employ a particular machine.

DVR principle

Figure 8.1 illustrates the operating principle of a digital video recorder. For the machine to be able to perform effectively, the core PC section should incorporate a high-speed dual-core processor, a large RAM capacity and a dedicated hard disk drive onto which will be installed the operating system, the DVR (and MUX) application software, drivers, etc. To increase speed and reliability, a separate hard drive (or drives) should be employed for the actual video storage. It is difficult in a textbook such as this to specify actual processor speeds and RAM capacities because technology in this area is progressing so rapidly that the text soon becomes out of date. As a guide, remember that video signal processing places a high

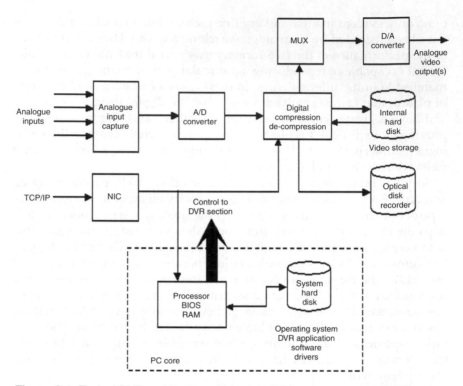

Figure 8.1 *Typical DVR architecture, based on PC hardware and operating system*

demand on a CPU, and you cannot have too much RAM capacity available for buffering purposes during compression/de-compression of the video images. So, for the next few years at least, it will be safe to say that the fastest processor available will be the preferred choice for any DVR, and RAM capacities should be measured in Gbytes and not Mbytes.

The analogue input capture card shown in Figure 8.1 plays a key role and may determine the performance of the DVR in terms of picture update rate and image quality. The capture card will have a number of analogue inputs ranging (typically) from 4 to 16. The speed and efficiency of the capture engines will determine the rate at which the functions of A/D conversion and digital compression are performed, and therefore the rate at which pictures are recorded.

In addition to the analogue inputs, the DVR will have a network connection (network interface card, or NIC) via which the machine may stream images from IP cameras and/or video servers that will have converted analogue camera signals into TCP/IP. The network interface is also used to provide remote access to the DVR from other PCs running either browser or administration software. Such access can provide remote configuration or reconfiguration, browsing of recorded images from a remote site, and control of PTZ cameras.

The hard disk used for video storage may be a single internal drive; however, this will have limited recording capacity, and offers no protection against loss of all video information in the event of catastrophic drive failure. Many machines offer the facility of removable, replaceable disk drives, and the larger machines are generally able to support RAID arrays (discussed later in this chapter).

An analogue video output is usually available, often in the form of both composite and S-Video. This output is normally used to provide a monitor facility, although it is sometimes used to download video onto a VCR or other external recording device. This output is normally provided via an internal MUX.

The optical disk recorder shown in Figure 8.1 is not necessarily a standard feature, but has been included to represent the fact that the machine must provide some means of extracting video information for the purposes of evidence or possibly archiving. Programmable options are often available to select automatic archive routines. For example, following an external alarm input activation, the unit automatically copies to DVD the data relating to the activation, beginning from a few minutes prior to the alarm.

The exact form that the archive device takes varies between machines, from a standard DVD recorder using MPEG-2, to some form of data disk recorder. However, it should be noted that in many cases the disks, including DVD recordings, cannot be replayed on the standard equipment incorporated in a PC without the special viewing software. This software is provided by the manufacturer and is often free, but it may not be immediately available in every police station or courtroom, which is where evidential material needs to be viewed.

Effects of compression

Without video compression, DVRs would not be viable because uncompressed image file sizes are very large, and even hard disk drives having capacities measured in terabytes would only provide very limited recording times. The principles and effects of video compression are discussed in detail in Chapter 5, but in this chapter, as we consider the DVR, we can appreciate the practical implications of compression in CCTV.

On the one hand there is a desire to compress the video signals as much as possible in order to extract the maximum recording time from the DVR, and yet on the other hand we know that excessive compression results in unrecoverable losses in picture information, leading to poor-quality picture reproduction. The ideal situation is one where we apply sufficient levels of compression to deliver a high recording capacity, but are able to extract high-resolution images from the machine. Unfortunately it is sometimes difficult to reach this position without spending a lot of money on multiple DVRs, especially in systems that have a large number of cameras and require a long archive period.

For many years the two principal compression methods employed in CCTV DVRs have been MPEG-2 and Wavelet (now JPEG 2000); however, since the implementation of MPEG-4, many machines are turning to this compression format as it generally offers improved image quality over MPEG-2.

Recording capacity

In general, hard disk video recorders are set up such that they will continue recording until the disk is full and will then proceed to overwrite the earliest recordings, thus providing continuous recording. Of course, problems will arise when the disk capacity is insufficient to provide the required archive period and thus the installer must ensure that the intended equipment is up to the job. This is more difficult for disk-based recorders than it is for VCR installations because, for VCRs all that is required is to look at the recording time (i.e., 12 hours) and ensure that the owner has enough tapes to cover the archive period. For example, for an archive period of 31 days, VCRs operating in the 12-hour time-lapse mode will require 62 tapes per machine plus, say, 5% to cover for tape extractions for evidential purposes.

In the case of hard disk recorders, the recording time is dependent upon a number of factors. These are illustrated in Figure 8.2. First of all there is

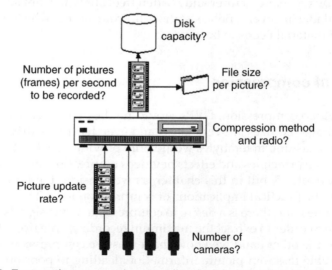

Figure 8.2 *Factors that govern the recording (archive) time for a DVR*

the *picture recording rate*. This is equivalent to the time-lapse period in a VCR where, on many DVRs, the installer or user can set the number of pictures per second that they wish to record. Secondly, there is the file size for each picture that is recorded, which is a product of the type and amount of compression that is applied. Because the images are digitized and compressed prior to recording, each picture in effect becomes like an individual file. The amount of compression can be set by the installer, or possibly by the user. The more compression that is applied, the smaller will be the file size and thus the greater will be the recording time. However, increasing the compression will reduce the image quality.

Another factor which determines the recording time is the *picture update rate*, which is not to be confused with the picture recording rate. The recording rate is the number of pictures per second (PPS) that the machine is recording, whereas the picture update rate is the rate at which each camera image is updated when reviewing the recording.

The number of cameras connected to the DVR affects the maximum recording time: the greater the number of cameras, the shorter the total archive period, assuming that the picture update rate is maintained. Finally, the recording time is determined by the capacity of the hard disk.

The update rate (in seconds) can be determined by dividing the number of cameras by the recording rate. Or:

$$\text{Pic update (in seconds)} = n \div \text{PPS}$$

where n is number of cameras, and PPS is the recording rate.

For example, for a system having just one camera, if the recording rate is set to 25 PPS, then the update rate will be $1 : 25 = 40 \, \text{ms}$. In other words, the replayed images for the camera will be updated every 40 ms, which is equal to the TV frame rate – that is, real time 25 pictures per second. If 25 cameras are now connected to the recorder and the record rate is maintained at 25 PPS, then the update rate becomes $25 \div 25 = 1$ second. So now we see that when the recording is replayed, the images from each camera will only be updated once per second.

Remembering that the PPS is a variable function of the DVR that is set up by either the installer or the user, the choice of setting for any DVR will be determined by two factors: the number of cameras connected to the recorder, and the rate at which the user requires the replayed images to be updated. Let's consider a more extreme example. A DVR is required to record the information from 30 cameras. The nature of the security risk requires a picture update of at least twice per second. Thus, from the expression Pic update $= n \div \text{PPS}$:

$$\begin{aligned} \text{PPS} &= n \div \text{Pic update} \\ &= 30 \div 2 = 15 \, \text{PPS} \end{aligned}$$

Thus, if the DVR's PPS setting is made to be 15, then it will meet the performance requirement for that system. However, this calculation does not take into account the archive period, and if the hard disk capacity is insufficient, the DVR will begin to overwrite video information too soon. Thus, to ensure that a DVR will provide a required archive period, the installer must calculate the hard disk capacity. In practice, many DVRs incorporate an internal calculation utility that will tell you the archive time for any combination of PPS, picture update and compression settings, but in order to use this you will have to have purchased the machine in the first place! Being aware of this, many of the leading DVR manufacturers offer calculation utilities on their websites, and/or through their technical support.

Using a longhand method, the required disk capacity for a system can be calculated from:

$$\text{Disk capacity} = 86\,400 \times \text{days} \times \text{PPS} \times \text{file size}$$

where 86 400 is the number of seconds in one day, 'days' is the number of archive days required, PPS is the number of pictures per second to be recorded, and file size is the size of each picture file (in kilobytes).

Note that the above expression and the examples given below assume that the recorder has been set to record the full frame size. Clearly, if the DVR has been set to record a reduced frame size, then there will be a reduction in the file size for each picture, resulting in a corresponding decrease in required disk capacity.

For example: a CCTV system requires an archive time of 14 days. It has been decided that the DVR will operate at a rate of 5 PPS and, for the proposed machine, the file size must be 20 kB per picture in order to obtain an image quality comparable to that of an S-VHS recorder. Thus, the DVR would require a hard disk capacity of at least

$$86\,400 \times 14 \times 5 \times 20\,000 = 121 \text{ gigabytes (GB)}$$

This example, which is typical of current digital recording equipment, illustrates just how much disk space is required. And yet the figures used in our example really only relate to a smaller installation because of the low picture recording rate (PPS). Let's now consider the larger system that we looked at earlier. This used 30 cameras and required a picture update rate of twice per second, which led to a required PPS of 15. Let's now say that the archive period for this system has to be 31 days. How much hard disk capacity would be required if the recording quality were to be similar to that of S-VHS?

$$86\,400 \times 31 \times 15 \times 20\,000 = 803 \text{ gigabytes (GB)}$$

If this system were expanded to 60 cameras, then the hard disk capacity would have to be increased to

$$\text{PPS} = n \div \text{Pic update}$$
$$= 60 \div 2 = 30 \text{ PPS}$$
$$\text{Disk capacity} = 86\,400 \times 31 \times 30 \times 20\,000$$
$$= 1.6 \text{ terabytes (TB)}$$

This capacity could be reduced by reducing the file size for each image; however, this will result in a reduction in image quality.

One of the prime factors in the rapid shift from VHS to DVR in CCTV recording was the availability of reliable, high-capacity hard drives. However, for many years such high capacities were only available by employing SCSI technology, which enables a number of hard disk drives to be daisy-chained in such a way that they behave as a single drive. DVR manufacturers took different approaches to using SCSI: some machines incorporated a number of internal SCSI drives, whereas others had external connections which enabled the installer to add drives as necessary. The problem with SCSI is that, although the drive speed is very fast, there is no protection against loss of data in the event of any one of the hard disk drives failing. When a disk drive fails, all data in the entire array is lost just as with a single drive, except that the chances of a drive failure are increased because there are more drives to fail.

RAID disk recording

An alternative to SCSI is RAID technology (redundant arrays of independent disks – although the original 'I' term was 'inexpensive') where a number of hard disk drives are grouped into an array, and the array acts as a single drive. Although this might sound very much like SCSI, the difference is that, with RAID, it is possible to build in protection against data loss in cases of a disk failure.

There are different levels of RAID; however, the principle is set out in RAID 0, which is illustrated in the left-hand array of three disks depicted in Figure 8.3. Each disk is referred to as a *column*, and each column is broken down into *stripes*. The *stripe width* refers to the amount of data in each stripe which, in the illustration in Figure 8.3, is 128 kB.

Now, imagine that we are recording a 1 MB file which, taking 20 kB as a typical file size for one compressed CCTV image, is equal to about 50 images. The first 128 kB will be written to column 1, stripe 1, the next 128 kB will be written to column 2, stripe 1, the third 128 kB to column 3, stripe 1, the fourth to column 1, stripe 2, and so on until all of the file has been written (which would be on column 2, stripe 3). Because of the parallel action of the disks, the writing speed can far exceed that of SCSI; however, RAID 0 does not provide any immunity to data loss in the event of a drive failure. Because the files are scattered across the disks, the loss of a disk means that the system would have no way of recovering that information. Therefore all of the data is in effect lost.

RAID 1 is, on its own, not really a RAID system at all; it is simply a specification for arranging backup disk drives to mirror the main drives. In RAID 1, as data is being written to each main drive, it is also being written to a separate mirror (backup) drive. In the event of a main drive failure, the mirror simply takes over and continues to run with the other main drives. When the defective drive is replaced, the data on the mirror is copied over to the new drive, which then continues to work in the main array.

The robustness of RAID only comes into play when we combine RAID 0 and RAID 1 to create a RAID 1 + 0 array. This is depicted in Figure 8.3 when we combine the three-disk RAID 0 array with the three mirror

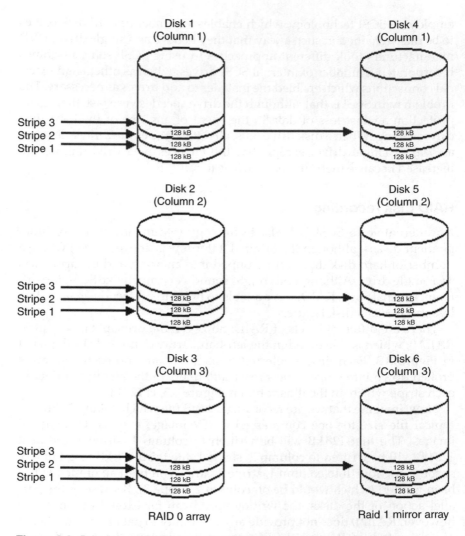

Figure 8.3 *Principle of RAID 1+0. The RAID 0 array (disks 1–3) has no protection against drive failure. When used in conjunction with the RAID 1 mirror array (disks 4–6), we have a RAID 1+0, i.e., a striped mirrored array*

drives. This arrangement operates very fast and offers high data reliability in the event of a drive failure. However, the cost and physical size is considerable when compared with SCSI because of the need for the mirror drives.

A less expensive alternative to RAID 1+0 is RAID 5, which is illustrated in Figure 8.4. Again we shall take the example of a 1 MB file being written into a series of 128 kB stripes. The first 128 kB block is written to columns (disks) 1 and 2. The second block is written to columns 2 and 3, the third to columns 3 and 1, and so on. In addition to each block of data, a parity

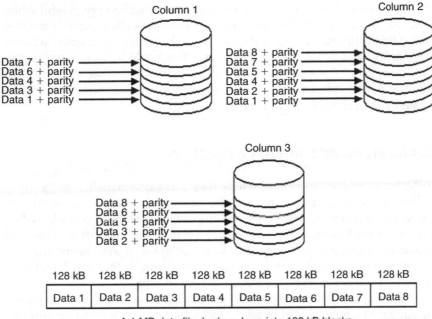

Column 1

Column 2

Data 7 + parity
Data 6 + parity
Data 4 + parity
Data 3 + parity
Data 1 + parity

Data 8 + parity
Data 7 + parity
Data 5 + parity
Data 4 + parity
Data 2 + parity
Data 1 + parity

Column 3

Data 8 + parity
Data 6 + parity
Data 5 + parlty
Data 3 + parity
Data 2 + parity

128 kB	128 kB	128 kB	128 kB	128 kB	128 kB	128 kB	128 kB
Data 1	Data 2	Data 3	Data 4	Data 5	Data 6	Data 7	Data 8

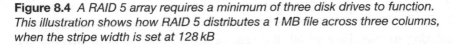

A 1 MB data file, broken down into 128 kB blocks

Figure 8.4 *A RAID 5 array requires a minimum of three disk drives to function. This illustration shows how RAID 5 distributes a 1 MB file across three columns, when the stripe width is set at 128 kB*

check value is recorded, although the data block and the corresponding parity are not recorded onto the same disk.

Now, let's assume that disk number two (column 2) has failed. No data is actually lost because it has all been duplicated on the other two disks. The system will continue to record data and parity (although now the parity will have to co-exist with the data) until a replacement drive is fitted. The system will then reconstruct the stripes on the new drive.

The one disadvantage of RAID 5 when compared with SCSI is that RAID 5 is slower due to the amount of duplicity in writing, plus the need to write the parity information. However, it is the preferred choice of many DVR manufacturers because its speed is sufficient for recording CCTV images, especially as disk drive speeds have increased in more recent years.

Many DVRs are sold as 'RAID ready', which usually means that they come with a single internal hard disk for use with smaller CCTV applications, but have a series of RAID 5 caddies into which the installer may add two or more drives to build up an array (bear in mind that RAID 5 requires a minimum of three drives to function).

Where a RAID 5 drive has failed, the DVR will put out some form of alarm to inform the user that there is a problem. It will also tell the engineer which drive has failed. However, the engineer must be very careful when replacing the drive that they pull out the correct one. Bearing in mind that the system is limping along with only two other drives (assuming a three-drive array) and is not able to construct a true RAID 5 block + parity pattern, removal of the wrong drive in effect means that a second drive has failed and at that point it is almost certain that all data will be lost. This action by the engineer is unlikely to amuse the user!

Digital video information extraction

DVRs have evolved to a point where they can easily match, if not exceed, S-VHS in terms of image quality and resolution (when comparing full frame size recording). However, one important feature to consider when selecting a DVR is its ability to produce a removable copy of the recorded information that is equal in quality to the original, is in a format that can easily be viewed, and contains a secure watermark that can be used to prove that the material has not been tampered with.

With tapes this was simple, because the original tape could be taken from the source machine, sealed in an evidence bag if necessary and could easily be replayed on any other S-VHS machine via a suitable de-multiplexer. The simplest way to replicate this process is to employ a DVR that has a removable, replaceable hard disk. However, there are a number of issues relating to this. First of all, the cost of a replacement hard drive is far greater than that of an S-VHS tape. Then, once removed, the drive will only function again when connected to a similar model DVR. Finally, what if the original material is spread across a nine-drive RAID 5 array? All nine drives would have to be removed and replaced.

In reality, in the UK, where a DVR contains (or is even suspected of containing) evidence relating to a serious crime, the police have the option to simply remove the entire DVR, possibly for many months. The problem with this is that in most cases the user has lost not only their recording equipment, but their entire MUX as well, which renders their system useless. An example of this, perhaps extreme but real nonetheless, is the action that police in London took in the immediate aftermath of the terrorist bombs on 7 July 2005. Many thousands of VHS tapes were removed from machines; however, these systems were soon up and running again because all that was required was a replacement tape. Unfortunately, systems that employed DVRs proved to be much more of a problem because many (not all, it must be stressed) did not have an acceptable archive facility, so the hard disk drive (and in some cases the entire DVR) was removed. Larger systems that employed a multiple of DVRs all using RAID disks suffered another problem: although archive was possible, it would have taken many hours for this to take place, during which time the machines would continue overwriting the oldest files. This was not acceptable to the police authority whose job it is to secure all evidence as soon as possible. So, once again in some cases all DVRs had to be immediately powered down and removed from site.

The above example is of an extreme nature because we are dealing with a very serious crime, at a level far greater than the average CCTV user is attempting to deal with when having their CCTV system specified. Consequently it need not be a requirement of every DVR to have some form of hard disk replication or mirror facility, and many machines do offer an effective archive utility. What the installer/specifier must do is ensure that the facility will meet the expected requirements of the system.

One common archive method is for the DVR to have an in-built DVD recorder. Video information that is deemed to be important may be copied to a DVD disk (usually in data not MPEG-2 video format) and removed. Many DVRs have the ability to automatically archive video to the disk whenever an alarm activation occurs. The problem with some of these disks is that they can only be played on a PC that is running suitable application software containing the codec, which is not always convenient. To avoid this problem, many machines that utilize DVD archiving make use of standard browser or media player software applications, or include the reproduction software on the disk.

DVD archiving is often acceptable for smaller systems; however, what options are there for larger recording systems? Where there is a requirement to store large amounts of data, perhaps in the gigabyte arena, perhaps the most obvious choice is to use one of a number of data tape backup systems which are capable of storing many hundreds of gigabytes of data on one single tape. However, where there is a chance that the information stored on those tapes may be required for evidence in a court of law, the tapes must be logged and stored according to control room standards, similarly to analogue videotapes.

Another option for large systems is for the user not to employ any recording equipment at all and, instead, have all video streamed from their control room to a third party provider. This third party will have their own secure recording facility amounting to many terabytes of capacity and, for increased security, all video data will be recorded in tandem in two separate locations. Using secure passwords and private fibre-optic connections, the user may still access their video information just as if it were stored locally. The advantages of this system are that the user is not troubled with maintaining their recording equipment and keeping it up to date, they still have access to all of their video content, and when an event occurs on a scale of a terrorist attack, there would be ample time for all data to be archived because the main servers could be prevented from overwriting, but could continue recording using their spare capacity.

At the time of writing this option is being considered by a number of larger CCTV system operators in the UK, and at least one large city centre CCTV system has taken this up using BT (British Telecom) services for both signal transmission and video storage. Time will tell how this progresses.

VHS recording

In order to record high frequencies on a magnetic tape, a very high tape transport speed is required. In all analogue video recorders this is achieved by wrapping the tape around a head drum and rotating the drum at a rate of 25 revolutions per second (i.e., a rate that is equal to one TV frame period). The head drum is tilted with respect to the tape so that the recording heads scan the tape following a diagonal path up the tape. Two head cores are fixed onto the drum 180° apart so that there is always at least one head in contact with the tape at any time. The principle is shown in Figure 8.5.

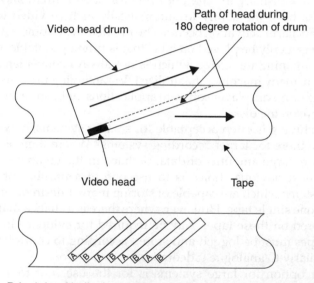

Figure 8.5 *Principle of helical scanning. Magnetic information is laid down on the tape in the form of diagonal stripes. This is known as helical scanning*

From this illustration it can be seen that:

- head A records the first TV field;
- head B records the second TV field;
- each single 360° revolution of the head drum records one TV frame.

During replay the machine requires a reference signal to synchronize the drum rotation with the tape transport. This is what is meant when we refer to *tracking*. The reference signal is provided from a control (CTL) track which is derived from the 50 Hz field sync, having first been divided down to 25 Hz before being recorded along the bottom edge of the tape (for NTSC, the 60 Hz field sync produces a 30 Hz CTL track). The CTL track is also used as a reference for the real time tape counter, and to mark the positions of alarm input activations in time-lapse VCRs to aid fast searching for the alarm events.

A videotape, therefore, contains the diagonal video signal tracks positioned in its centre, and lateral audio and CTL tracks along the edges. This track format is shown in Figure 8.6.

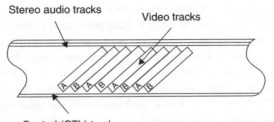

Figure 8.6 *Videotape track format*

Mistracking occurs in a VCR when, during replay, the tape transport and drum rotation motors are both running at the correct speed, but they are not in phase with each other. In this case the heads will be scanning two tracks at the same time, resulting in a low output signal from the playback heads.

Mistracking is a result of small mechanical tolerances between machines and, to compensate for this, manufacturers include a tracking control circuit. Tracking problems can also occur when a tape has become stretched. This alters the angle of the video tracks, making it difficult for the heads to follow them. Similarly the machine may suffer mechanical wear, making tracking difficult. Where tracking errors are caused by faulty tapes or machines it is often difficult to correct using the tracking control.

VHS is only capable of recording luma signals up to 2.8 MHz, which corresponds to a resolution of around 240 TVL. An S-VHS machine is capable of reproducing video signal frequencies up to 5 MHz, which equates to a TVL in the order of 400.

It is possible to record in standard VHS on a S-VHS tape, but it is not possible to record in Super format using a standard VHS tape because of the limitations in the tape oxide. To prevent this from happening by accident, manufacturers fit a sensor in the machine which detects a S-VHS tape. If a standard format cassette is inserted into the machine, then even if the record selection is set for Super format, the machine will revert to standard format recording. Engineers should make customers aware of this, pointing out that they are compromising the quality of the system if they use the much less expensive standard cassettes.

Time-lapse recording

By making use of the still frame function, in the recording mode time-lapse machines record a single field, pause momentarily, and then record another field. Whilst the machine is in the pause mode the head drum continues to rotate; however, the heads are switched off. Note that because the tape stops regularly, it is not possible to record an audio track whilst the machine is operating in time-lapse mode.

By operating in time-lapse mode a three-hour cassette can be made to last much longer. Of course the tape only contains a series of still images, but this is often sufficient for CCTV purposes, and if the tape is replayed in a faster mode than that in which it was recorded, a form of fast motion film can be viewed. The number of still pictures available on one three-hour cassette is quite impressive. In the three-hour mode a machine is recording 25 pictures per second, and thus over three hours this amounts to a total of $25 \times 60 \times 60 \times 3 = 270\,000$ pictures. In time-lapse mode each of these frames is an individual picture, and thus the cassette may be considered to be like a photograph album containing 270 000 photos.

The method of recording one TV frame per time-lapse interval is in fact very wasteful because, when a VCR replays a still frame it actually only scans one field of the frame, and the other field is effectively ignored. This point is illustrated in Figure 8.7. Therefore it was decided that, if the second field of each frame is ignored, then why bother recording it? Consequently, time-lapse VHS machines actually only record one TV field out of each frame, meaning that we can now record twice as many pictures, i.e., $2 \times 270\,000 = 540\,000$ still fields/pictures.

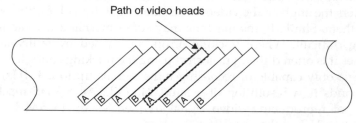

Path of video heads

Stationary tape

Figure 8.7 *In still mode the tape is stationary, and thus both video heads scan the same track*

An important consideration with time-lapse operation is the loss of vertical picture resolution. In Chapter 5 we saw that a TV picture contains at the very best 625 lines less 40 vertical flyback lines, equalling 585 active lines. However, when we consider that these lines are contained within two fields, clearly when a VCR is in still mode, although it is still producing two fields of 312.5 lines, they are in fact the same lines. In other words, the vertical resolution is reduced by 50%. This loss may be crucial when a recorded picture is to be used as evidence.

Not all machines have the same number of time-lapse modes, although the more elaborate models offer a very wide range. Some typical modes, along with the delay periods for both frame and field operation, are given in Table 8.1.

Table 8.1 *Time-lapse modes. Note that the actual recoding period is always three hours longer than the stated period. This is due to the way that VHS time-lapse recorders have to arrange their video tracks*

Mode	Running time	Delay
12 h	15 h	0.10 s
24 h	27 h	0.18 s
48 h	51 h	0.34 s
72 h	75 h	0.50 s
120 h	123 h	0.82 s
240 h	243 h	1.62 s
480 h	483 h	3.22 s
720 h	723 h	4.82 s
960 h	963 h	6.42 s

Linear slow speed is a form of 'continuous' time-lapse where 12 or 24 hour recording is achieved not by pausing the tape but by running it continuously at a speed that is four or eight times slower than normal. This was introduced solely to enable audio recording. Because the tape is running continuously, a lateral sound track can be laid down at the top of the tape, albeit of very poor quality owing to the slow tape speed.

When *alarm mode* is selected the machine operates in time-lapse mode, but when the alarm input terminal on the machine is active, the machine reverts to the three-hour mode for a preset period (usually a few minutes), thus ensuring that maximum information is recorded. The alarm input may be triggered manually by an operator using a simple push switch, or it could be connected to some form of electronic alarm system. On a larger system where the machine is expected to record the output from a number of cameras, the alarm activation signal can also be connected to the multiplexer or switcher. The normal recording pattern can then be interrupted so that only the cameras in the activation area are connected to the VCR input during the alarm period.

Auto re-record is a feature which is particularly useful for unmanned systems. As the name implies, when the tape comes to the end, the machine rewinds and re-records. Options are usually built into this feature to prevent erasure of important evidence. For example, the machine may be set so that it will not auto re-record when an alarm input activation has taken place.

VCR maintenance

Perhaps the biggest criticism of VHS when employed in the CCTV industry is the need for mechanical maintenance. And this criticism is justified because mechanical parts in a VCR deck do wear out with constant use. Numerous CCTV engineers can recite stories of customers who have invested thousands of pounds in CCTV equipment, only to penny pinch on tape replacement, and it is not uncommon for a customer to complain to an installer that, following an incident, the tape has been replayed only to produce a snowstorm. Upon inspection it is often found that the tape in question is the one which was provided during the system handover a number of years previously! In such circumstances, not only will the tape be next to useless (the oxide coating will be thin, and the backing material stretched) but the machine will desperately require a major mechanical overhaul before it is capable of reproducing anything like an acceptable picture.

In practice, a machine that has been operating continually for twelve months will be in need of a service. Even if the picture quality still appears to be satisfactory, there is no telling how much longer this will be the case.

When considering VCR servicing, video head drum checking and replacement is often singled out as the primary concern, yet this is not the only item that may be showing signs of wear after a machine that has been operated continuously for one full year (8760 hours). Items such as the pinch roller, rubber belts, back tension band (if used) and head cleaning rollers will all require checking and/or replacement, and the condition of the lower video drum, tape guide posts and audio/control head should be verified. Furthermore, the tape path must be cleaned of all dirt and tape oxide deposits, and the alignment of tape guides checked and adjusted as necessary following video head drum replacement. In some cases, manufacturers provide service kits containing all of the mechanical components which should be replaced after a period of no more than 10 000 hours. Clearly the level of servicing being discussed here can only be performed by competent VCR servicing personnel, and for this reason it is expected that a machine will be removed from site and returned for (annual) servicing.

Video head cleaning

The frequency at which video heads require cleaning can be a matter of opinion, and in reality is dependent upon a number of factors including the quality of tapes being used, the number of times a tape is to be used, the tape storage method and conditions, whether the machine has an automatic cleaning facility, and a degree of luck! But before we consider the frequency and methods of head cleaning it is important to understand what a 'dirty head' is, and recognize the symptoms.

The 'dirt' we refer to is not the airborne dust that we see collecting on monitor screens and cabinets. It is particles of tape oxide which have detached from the tape and have either adhered to the side of the video head drum or become lodged in the head tip. This is illustrated in Figure 8.8. Oxide on the side of the drum reduces the tape/head contact, which in turn reduces the record/replay signal levels. The overall effect is poor tracking symptoms. When the oxide is lodged in one of the heads, the effect on the picture is the same as if one video head had failed.

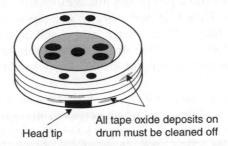

Head tip

All tape oxide deposits on drum must be cleaned off

Figure 8.8 *Typical oxide build-up on a VHS video head drum*

Head cleaning is the removal of the oxide from both the head gap and drum surface. However, this is not as simple as it might first appear, because the video head tips are extremely fragile, and the slightest pressure in the vertical direction will break them off, necessitating replacement of the entire drum assembly. Oxide on the drum surface can become well-adhered, and there is a temptation to try to scrape this off. However, the smallest scratch on the drum surface will alter the aerodynamics to such an extent that the tape will ride off the drum, producing a permanent line across the picture.

Head cleaning cassettes are available, and some manufacturers supply one of these with each time-lapse machine. During handover the customer should be instructed in the use of these, but should be warned that excessive use may lead to early head wear, as some of these can be somewhat abrasive. As a general rule, if the picture is perfectly all right, don't bother cleaning the heads!

Some machines employ an automatic head-cleaning facility. This usually operates on the principle of a felt wheel that is pressed against the head drum each time the tape loads and unloads. Where this is the case, the customer can be instructed to load and unload a tape a number of times to force this cleaning operation in the event of apparent dirty head symptoms.

The problem with both cleaning cassettes and automated systems is that, when the oxide is lodged into the gap or fixed firmly onto the drum, there is insufficient pressure to dislodge it. In this case the only option is to clean the head drum manually, and because of the delicate nature of the head drum, this operation should be performed only by those who have been trained in the technique.

Tape management and care

Although this is primarily the responsibility of the operator, the engineer should make their customers aware of the implications relating to tapes during the initial system handover.

A tape passing through a machine operating in the time-lapse mode is subjected to far more stress and wear than when it is used at normal speed because each part of the tape spends much more time stationary against the rotating heads. Thus it is necessary to use tapes that have a strong backing material which will not stretch, and a well-bonded oxide coating. Not all VHS cassettes are of the same quality in this respect, and this is why only tapes marked as being of 'professional' quality should be used for CCTV applications.

Another reason for using quality tapes is that not all oxide coatings have the same magnetic qualities, the budget versions often suffering from high-frequency losses and lower signal-output levels (poorer S/N ratio).

The number of times that a tape should be used is a matter of some contention, but experience in the industry indicates that for tapes being used in the 24-hour time-lapse mode, twelve passes through the machine is the maximum. Thus, if a customer is going to operate the machine in the 24-hour mode, purchasing 31 tapes will allow a suitable rotation over a twelve-month period. However, this does not allow for any losses incurred when tapes are subsequently removed from the system because they contain evidence.

The Data Protection Act 1998 requires that a detailed log of all CCTV tape movement is maintained, and the engineer should encourage the customer to use an auditable tape logging and management system. For smaller installations it is often possible to make use of systems that are provided with some of the videocassettes which are manufactured specifically for CCTV use. However, where a manned control room is in operation, a more sophisticated tape management system will be required. These may be purchased ready made, and include a secure storage cabinet which is designed to aid tape rotation, and a complete set of logging sheets.

With regard to tape storage, advise the system owner to store tapes flat and not vertical, to have a storage environment that is not too hot or too cold (tapes are designed to live in 'normal' room temperatures, just like ourselves!), and to ensure that that there are no stray magnetic fields in the area which will cause at least some degree of erasure: video monitors leak magnetic fields from all sides, so keep tapes well away.

Digital video tape

In addition to the poor resolution offered by analogue video recorders, another serious problem for the CCTV industry is the fact that the signal degrades every time the tape is replayed or a copy is made. To overcome all of these problems a number of digital video tape recording formats have been developed and, although digital video recording is by no means faultless (it suffers from digital noise and compression losses), it provides a clean, quality picture which does not degrade with either time or through generations of duplication.

One such format is the digital time-lapse VCR (D-TL) developed by Panasonic and Sanyo. This digital recorder employs a normal S-VHS tape cassette and in many respects functions as a traditional analogue machine. However, the digital signal processing means that a horizontal resolution of up to 520 TVL is possible. The D-TL combines the advantages of digital video recording with the relatively low cost of S-VHS tapes. Perhaps one disadvantage is the fact that the recordings can only be replayed on D-TL format machines meaning that, when data is required for evidential purposes, it will more than likely have to be copied down to S-VHS or even VHS. Nevertheless, because the original material is of a high quality in terms of resolution, this does mean that the copies will be of a high standard.

Another digital video tape recording format, known as 'DV', has been developed by Sony with both industrial and domestic applications in mind, and its resolution of up to 500 TVL makes it attractive to the CCTV industry.

There are two versions of the DV cassette: DV and Mini DV. Both cassettes use a 6.35 mm wide (1/4") tape, but the Mini DV cassette is much smaller, with a correspondingly shorter recording time. The Mini DV tape is used in the majority of domestic digital camcorders because it keeps the size of the machine to a minimum; however, for CCTV applications the 60-minute recording time is somewhat restrictive, and many archive machines take the full-size cassette which provides 4.5 hours of recording time. Some models have the facility to take either size of cassette.

9 Camera switching and multiplexing

On larger installations it is clearly impractical to have one monitor for each camera, and therefore some means of image selection must be employed. There are basically two options available: we can sequentially switch between cameras, or we can display a number of camera images on a single monitor screen. Both of these methods have their strengths and weaknesses. Switching implies that the operator is only able to watch a single picture at a time (although as we shall see later, more than one monitor may be used) and therefore each area has a period during which it is not being monitored. Multiple imaging perhaps removes the blind spots; however, where there are too many images on display, it becomes difficult for an operator to maintain the concentration required to look at all of these images, and the reduced size of the images often leads to a loss of definition.

Of course, these are very much generalizations, and developments in technology have done much to overcome the disadvantages, often by bringing the two concepts together.

Sequential switching

The principle behind video switching is shown in Figure 9.1, where simple sequential switching is performed by the electronic switch S_1. The rate at which the switch scans the inputs, known as the *dwell time*, would be set by the operator using a simple keypad input.

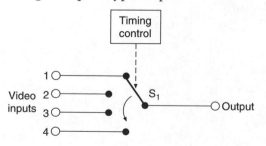

Figure 9.1 *Principle of sequential analogue switching*

A more practical sequential switcher is shown in Figure 9.2. This arrangement makes provision for alarm inputs. Under normal circumstances S_3 is closed and the timing control circuit operates S_1 and S_2 in tandem. Hence the video recording equipment is recording whatever the operator is viewing. Upon receipt of an alarm, for example on input 3, switch S_3 opens, S_2 is moved permanently to position 3 and the warning output is activated. In this condition the main monitor continues to display all four cameras sequentially; however, the operator can investigate the cause of the alarm on a spot monitor whilst recording the activity. The operator can be alerted to the alarm by a light and/or buzzer connected to the alarm output, which may also be used to activate the alarm input on a DVR/VCR (see Chapter 8).

Before proceeding it is worth reminding ourselves of the importance of output termination, which was discussed in Chapter 7 (see Figure 7.20). Any unused outputs should be correctly terminated using either the termination switch on the back of the equipment (if fitted), or a $75\,\Omega$ termination device. In all modern CCTV equipment switching is automatic, the $75\,\Omega$ resistor on an output being switched out of circuit when the BNC connector is pushed into place. The rules for termination apply to all of the equipment which we shall be looking at during the course of this chapter.

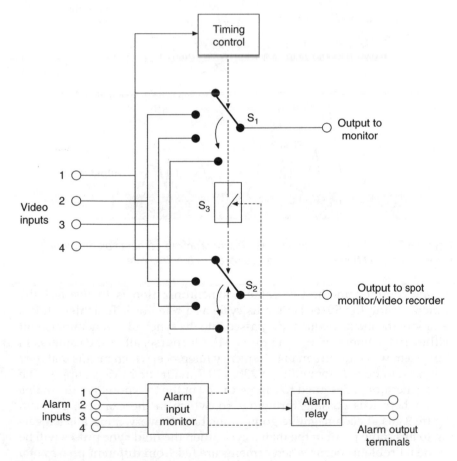

Figure 9.2 *Sequential switcher with alarm inputs*

Sequential video switchers have no means of maintaining synchronization between cameras and therefore, unless video signal synchronization is performed elsewhere such as in a DVR, the picture is likely to roll each time the switch toggles because of the difference in the scanning positions at each camera. This is illustrated in Figure 9.3, where the phase relationship between the video signals from two unsynchronized cameras is drawn as they would appear if they were displayed on a dual beam oscilloscope. From this illustration it can be seen that camera synchronization is a matter of ensuring that each camera begins scanning the first TV field at precisely the same instant. When this condition exists between every camera in an installation, there will be no picture roll when a switcher toggles between cameras because the monitors will not see any change in the timing of the vertical sync pulses.

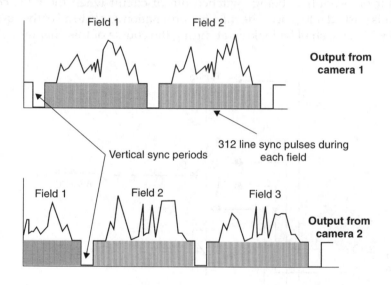

Figure 9.3 *Relationship between two unsynchronized cameras as they would appear on a dual beam oscilloscope adjusted to the field rate*

One common method of camera synchronization is to *line lock* the cameras using the 50/60 Hz mains cycle as a reference. To achieve this a sample of the mains frequency is passed into the sync pulse generator circuit within each camera (refer to Figure 6.11). Of course, all of the cameras in the system will require a mains frequency reference, which means that they will have to be fed from either a 230/120 V mains or 24 Va.c. supply. The sync generator is designed to trigger to one of the two points in the mains cycle where it is passing through zero, which in the example given in Figure 9.4 is at each negative going transition. If every camera is triggering to this same point in the mains cycle, then the field sync pulses will be aligned. Problems occur when cameras are fed from different phases of a three-phase mains supply and, to compensate for this, cameras with the line lock facility incorporate a *vertical phase control* which, when adjusted, alters the timing of the sync generator circuit. The adjustment normally has a range of at least 120°, which is the difference between any two phases in mains supply. This control may be set by trial and error; however, where this is done it is possible that some cameras are set 'just on the edge' and, when changes occur in the mains supply voltage, these cameras may manifest problems of picture roll during switching.

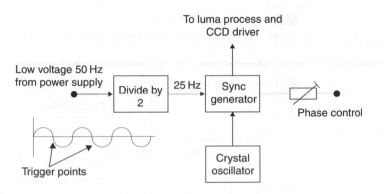

Figure 9.4 *Line lock arrangement in a camera*

A more sure method of adjusting camera phase is to use an oscilloscope to display the vertical sync phase as shown in Figure 9.3. One camera must be selected as a reference, and the output from every other camera is then set against this, the V phase control being adjusted until the vertical sync pulses are aligned. Of course, this operation is not as simple as it may sound because the oscilloscope may be located at the control room in order to access the camera inputs, and therefore the engineer will be unable to view the scope display from the camera location. Using two people with a mobile radio or mobile phone is one solution to the problem. A better solution is to take the oscilloscope out to the camera that is to be adjusted (a hand-held scope is clearly the most practical device for this operation) and connect the camera directly to the scope. The feed from the reference camera may then be provided by linking the cables from the two cameras at the control room. The principle is shown in Figure 9.5.

Another method of maintaining synchronization between cameras is to use *genlocking*. This is where a vertical sync pulse is fed to an input at each camera which is labelled 'genlock', 'ext sync', etc. (see Figure 6.11). Not all cameras have this facility, and it is therefore important that suitable models are chosen for installations which are to employ genlocking. There are two ways of deriving a master sync pulse. One is to take the video output from one camera and distribute this to all of the other cameras in the system. In this case the sync selector switch at the first camera would be set to the 'internal' position, and the switches on all subsequent cameras set to the 'ext sync' position. This method is illustrated in Figure 9.6.

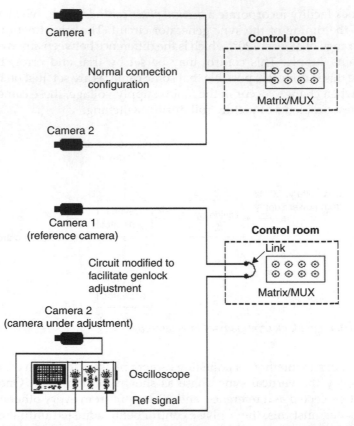

Figure 9.5 *Method for performing genlock. Both cables are disconnected from the matrix/MUX and are linked together at the control room. Hence, the cable for Camera 2 is utilized as a feed for the reference signal from Camera 1*

The second source of a sync pulse is to use a master sync generator which would be located in the control room, the sync signal being distributed around the system.

Genlocking is a very robust method of synchronization, but it requires the installation of two co-axial cables to each camera location, which increases the cost. The only way that this can be avoided is to employ camera/ switcher combinations which send a sync signal through the video signal cable and into the video output socket on the camera. Such systems tend to be self-contained in that you can only use the cameras and equipment that have been specially designed to operate together, which is not a problem as long as you are not intending to extend an existing conventional system, or are required to extend the system beyond its capacity in the future.

Genlocking works well in a television studio where all cameras are linked using custom multicore cables. However, because of all the extra cable installation required, it was never really viable for CCTV and is

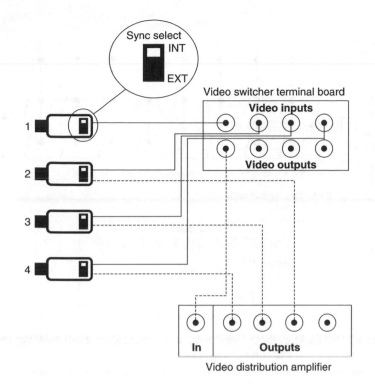

Figure 9.6 *One method of genlocking cameras. In this case all cameras are locked to the vertical sync pulses generated by Camera 1*

largely obsolete. Similarly, the need to line-lock cameras is now rare because the majority of DVRs are able to perform camera synchronization.

Matrix switching

Small-scale switching is generally performed in the DVR/NVRs. However, for large CCTV installations there is still a need for matrix switching. As the term matrix implies, this equipment is capable of selecting any one of a number of inputs and connecting it to one or more of a number of outputs. The principle is illustrated in Figure 9.7, which shows an 8 in 2 out matrix.

The electronic switches are controlled by the microcontroller chip. When the operator selects output 2, input 4, the switch corresponding to these co-ordinates is closed, and the signal from input 4 will be present on output 2. In this example the control console is simple and does not offer many features other than an alarm facility similar to that shown in Figure 9.2 (the camera input corresponding to the activated input being made available on output 2) and a sequential switching option on output 1.

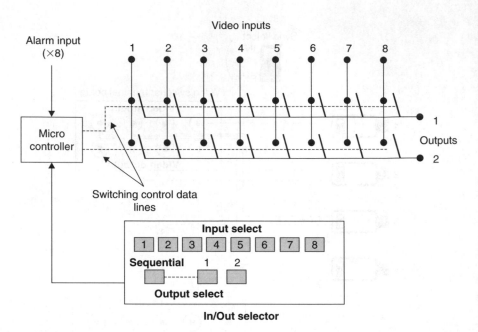

Figure 9.7 *Principle of matrix switching. In a practical, modern switcher the user interface is usually a PC software application*

One essential feature of a matrix switcher is that it can be designed to be expandable in terms of both inputs and outputs. Taking the example in Figure 9.7, if the manufacturer adopts a modular design, then the number of inputs can be increased to 16 by the insertion of an eight-channel input card identical to the first. Further cards may be added to give 24, 32, 40, 48, etc., inputs. Some larger units are capable of expanding up to many hundreds of inputs. Similarly, output expansion cards enable the number of available outputs to be increased. An expanded version of the unit in Figure 9.7 is shown in Figure 9.8.

In this arrangement the expansion card mimics the first input card (in this example an output expander has not been installed). The alarm input facility may not be required in all systems; therefore, the alarm expander may be separate from the video card so that it can be included as an option. Beyond 16 inputs the switch selector type shown in Figure 9.7 becomes impractical because of the number of buttons required and, for many years, a ten-digit keypad proved to be more practical, although modern switchers employ a PC-based software user interface.

The type of equipment shown in Figure 9.8 is a vast improvement on the sequential switcher; however, as the number of inputs is increased, certain limitations become apparent. Most noticeably, it would be difficult enough for a single operator to switch through 16 cameras, but to cope with any more would make the task impossible. This problem is overcome by enabling the system to have more than one control keyboard, and hence more than one operator. However, this creates another problem: how to

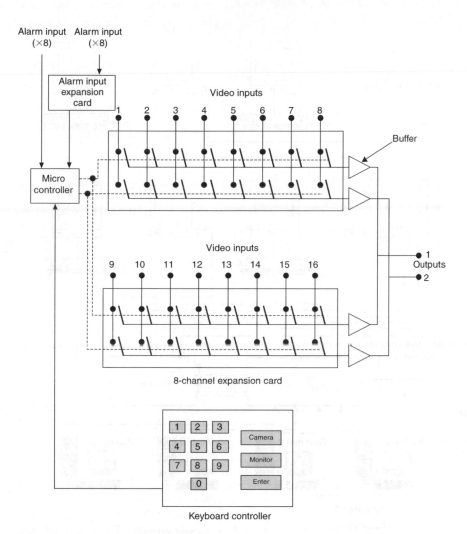

Figure 9.8 *An 8 channel matrix with one 8-channel expansion card installed*

prevent a number of operators from trying to access the same inputs at the same time. This can be taken care of in the system programming whereby each keyboard is given access to only certain cameras (inputs), i.e., each operator 'patrols' a given area within the CCTV system. Furthermore, these inputs will only be placed onto the monitors (outputs) associated with that operator, thus preventing one operator from bringing images onto another operator's screen. The programming can be further enhanced to allow two operators access to the same inputs; however, one of them will always have priority so that in the event of both attempting to access the same input, the operator with priority will always gain the access.

Another limitation of the simpler matrix systems is the inability to control the cameras. It is to be expected that a CCTV system with many dozens

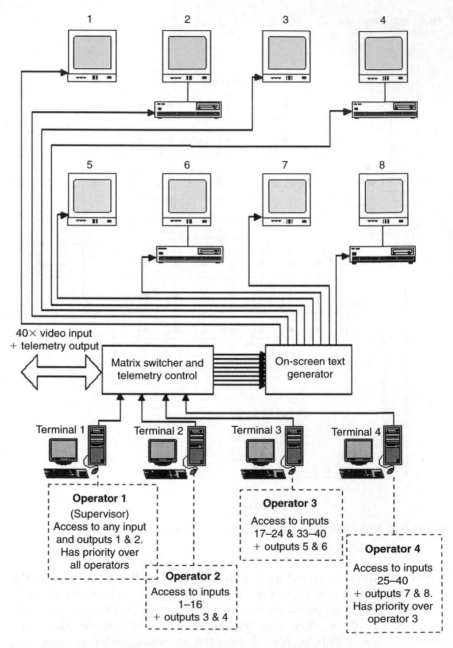

Figure 9.9 *A large matrix controller with multiple operator facility*

of cameras must require some of them to have a pan, tilt and zoom (PTZ) facility. This is why the larger matrix units have the capability to offer telemetry control. Because the number of PTZ outputs required will vary from system to system, the telemetry capacity is usually expandable in a similar way to the alarm input capability. Telemetry control will be discussed in more detail in Chapter 10.

A large matrix system illustrating many of the features we have been considering such as multiple operators, levels of operator access and telemetry control is shown in Figure 9.9. Note the inclusion of a DVR at each operator position, enabling incidents to be recorded whilst still being able to continue monitoring. Also note the inclusion of an on-screen text generator. This is essential for two reasons: in a large system it would be impossible for the operator to know what he/she was looking at without some form of camera and/or area identification facility, and in the UK it is a requirement for all video evidence presented in court to have a time and date displayed on the screen.

The problem with such a system is that, despite the sophistication of the switching control, each operator is still only able to view one scene at a time, and for a large number of cameras it is simply not possible for a person to take in a large number of changing images for any length of time. The problem can be alleviated to some degree by making use of the alarm inputs in such a way that each operator only scrolls through a small number of areas, selecting the others only if and when an alarm signal is received. However, reliance on an alarm detector may not be satisfactory for some situations, and the system could never be made to work in a town centre situation where all of the alarms would be triggering all of the time! It is clear that there are occasions when the operator needs to see a number of images simultaneously, and for this we need some form of screen splitting equipment.

The quad splitter

As the name implies, this equipment allows four camera images to be displayed on a screen simultaneously. In order to be able to achieve this, the incoming analogue signals must be digitized and stored in a frame store. Once in the store a central processor chip (CPU) has to reduce the size of each image to one quarter of the original, and arrange the data so that when it is clocked out and converted back into analogue form, the monitor will display four images in four quadrants on the screen. To do this, each horizontal line will contain video information for one picture during the first $26\,\mu s$, and for the adjacent picture during the second $26\,\mu s$ (remember that, for UK television, one active line period equals $52\,\mu s$). Likewise the top and bottom halves of each field contain information from different cameras. The process is illustrated in Figure 9.10.

The fact that each horizontal line now contains information which has been derived from two horizontal lines belonging to two different cameras implies that the horizontal sync pulses must have been removed by the quad unit, and a new sync pulse inserted. The same applies to the vertical sync pulse. This means that, because the quad is acting as a form of master sync generator (although it is not actually synchronizing the cameras

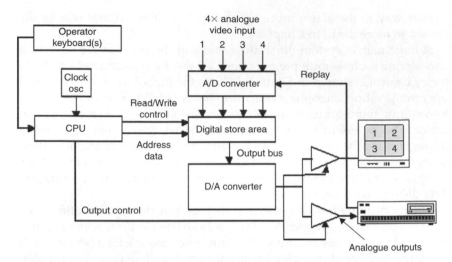

Figure 9.10 *Basic signal process in a quad splitter*

themselves) there is no requirement for the cameras to be synchronized by any other means. In other words, there is no requirement for line locking or genlocking of the cameras.

The problem with compressing a 52 μs line into 26 μs is that the horizontal resolution must be reduced, in theory by 50%. In basic quad switchers this is precisely what happens and so, although the operator can view four cameras simultaneously, it is at the expense of picture quality. Also, because it takes time to process the information, it is not possible to show four live action images; rather, the images move in a succession of frame jumps. To enable the operator to view high-resolution live images, the quad normally has the ability to display any selected input in normal full size, and alarm inputs may be included to prompt an operator to select this mode.

In the basic quad the problem of resolution loss becomes more acute when the signal is recorded onto videotape because there is no way of decompressing the compressed information. The implication is that, when the tape is replayed, the four images are permanently compressed, and even though the quad unit may have a VCR input facility which enables any one image in the replayed quad format to be separated and displayed at full screen size, the resolution is still only 50% of the original, both vertically and horizontally. In DVRs this is not usually a problem because each image will still be recorded at full resolution.

Earlier quad units were limited by the speed at which the CPU could manage the data process. If we examine this process in relation to Figure 9.10, we see that when the unit is operating in the quad mode the CPU has to clock four unsynchronized TV frames into the data store, strip out the line and field sync pulses, compress the data into 26 μs horizontal line periods, add new line and field sync pulses, and clock the data out of the frame store and into the D/A converter in the order that it is to appear on the screen. For this to be done in real time takes a considerable amount of processing speed and power, and the relatively low clock speeds of the early microchips meant they were simply not up to it.

Developments in microchip technology have brought about giant leaps in processor clock rates, and this has enabled quad splitters operating in real time to be developed. Some of these are capable of accommodating eight inputs. In normal quad mode the images are arranged into two groups of four, each group display being alternated on the screen, with a variable dwell time.

By increasing both the memory capacity of the digital store area and the CPU clock speed, the rate at which pictures are processed is increased, enabling a much faster rate of update. But perhaps the most significant improvement is in the fact that, although the resolution is initially lost when the images are converted to quad mode, because modern digital compression techniques are used (i.e., MPEG or Wavelet), the resolution can usually be recovered when the images are restored to full screen size.

Video multiplexers

The multiplexer (MUX) can be said to take over from where the quad leaves off. Taking advantage of the faster picture processing made available through high-speed CPUs and digital compression techniques, multiplexers offer a whole range of facilities that greatly enhance the effectiveness of a CCTV system. A typical multiplexer will offer a range of screen displays; some typical examples are shown in Figure 9.11. However, as we shall see later in this chapter, the multiplexer is not only capable of producing some clever screen displays; sophisticated models are capable of delivering two completely different picture sequences or structures to the monitor and the DVR, a facility which may be further enhanced when used in conjunction with alarm inputs.

The operating principle of the multiplexer is illustrated in Figure 9.12. Following the process through, the analogue composite video signals coming from the cameras are immediately converted into digital signals and stored in a temporary frame store area. Working at very a high speed, the video compression encoder reduces the amount of data per TV frame before placing this data into another store area. The video compression and expansion process is a complex operation requiring a lot of processor power, and to prevent the system from being slowed down, or even prone to lock-up (crashes), it is common for the compression circuits to have their own control CPU, freeing up the main CPU to deal with the multiplexing and other housekeeping operations. The compression circuits also have their own dynamic RAM (DRAM) memory stores to avoid clashes within the frame store areas.

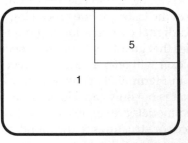

Figure 9.11 *Typical examples of screen layouts available with a multiplexer*

Multiplexing takes place as the pictures are moved out from the digital store area 1 by the main CPU, which clocks the data to the D/A converter in an order that will produce the desired picture layout on the monitor screen. When the operator alters the layout via the keyboard, the CPU simply changes the order in which the data is clocked out. Fast address and control buses are required between the CPU and the memory chips to sustain such complex data control.

Where a camera input signal is lost, the MUX will usually detect the loss of sync pulses and display a warning on the screen. When a sequential switching mode is selected, the MUX skips any missing or unused channels – a feature that is also found in many sequential switchers.

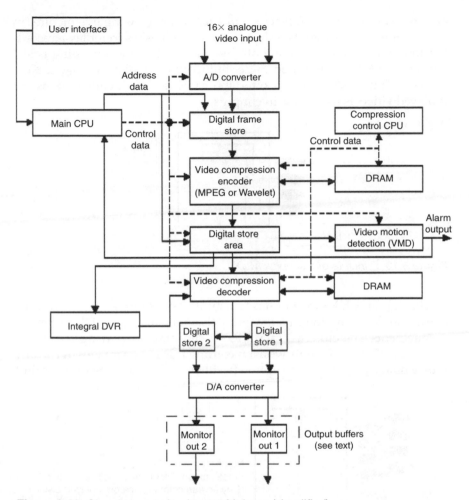

Figure 9.12 *Signal processing in a multiplexer (simplified)*

In units which have the provision for different monitor outputs, a second store area is required (Digital store 2 in Figure 9.12) so that data can be output to the D/A converter in different sequences for each output.

The main implication for VCR recording is that, unlike with the quad, the machine is able to record complete TV fields. Replaying a tape with this recording pattern directly into a monitor would produce unintelligible images; however, when the recording is replayed through the multiplexer, the images are able to be manipulated into whatever format the operator desires, i.e., full screen, quad, 9-way, etc. Note that in full-screen mode the movement may appear jerky because the VCR has not recorded every field, and therefore the multiplexer is effectively having to display a series of still frames.

Multiplexers are categorized as *simplex*, *duplex* or *triplex*. A simplex MUX has just one multiplexer (Figure 9.13), and therefore its capabilities are limited to recording full-screen images whilst viewing live pictures in a number of screen layouts. When the operator wishes to replay a recording through the simplex unit, both live monitoring and recording is lost because the MUX is required to de-multiplex the replayed images. Note that the unit is still able to offer the same screen layout options for replayed video as it does for live inputs.

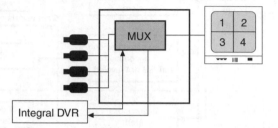

Figure 9.13 *Principle of a simplex MUX*

The duplex MUX incorporates two multiplexers (Figure 9.14), enabling one to handle the live information whilst the other takes care of the recording requirements. Therefore, a duplex MUX is capable of displaying live images in a range of screen layouts whilst recording full-screen images. It can also replay images without interruption to either the live viewing or recording.

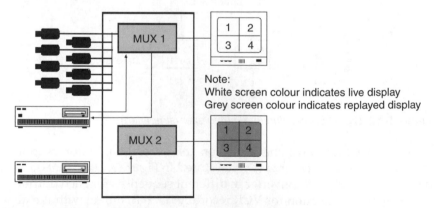

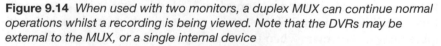

Figure 9.14 *When used with two monitors, a duplex MUX can continue normal operations whilst a recording is being viewed. Note that the DVRs may be external to the MUX, or a single internal device*

The triplex MUX principle is illustrated in Figure 9.15. The unit does everything that the duplex can do; however, it is possible to view live and recorded images simultaneously on the same monitor screen, without interruption to normal recording. The operator still has all of the usual options of picture enlargement or PIP. Alternatively, two monitors may be used to display live and recorded images in various screen layouts.

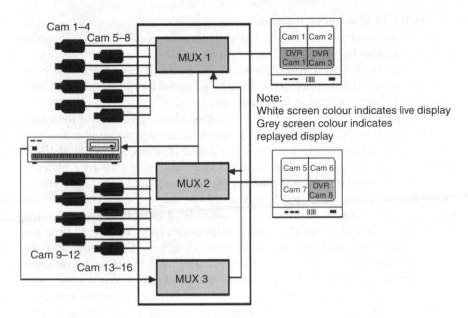

Figure 9.15 *A triplex MUX can display both live and recorded images on one screen*

It is not uncommon for an MUX unit to incorporate telemetry controllers; this subject will be covered in Chapter 10.

On-screen information is essential for identifying the date, time and location of video evidence, as well as assisting the operator in identifying an area quickly. The MUX usually has an *on-screen display (OSD)* facility, which is set up during installation.

Video motion detection (VMD)

Many CCTV installations rely on intruder alarm technology detectors such as passive infrared, infrared beam and microwave, to activate the alarm inputs; however, an alternative to this is to use the images coming from the video cameras as a means of intruder detection.

The principle of the VMD is to store a sample frame from the camera input and then compare subsequent frames with the sample, looking for changes in picture detail. Naturally it cannot look for a change in all of the picture information before triggering an alarm, otherwise the intruder would have to fill every part of the screen. Instead, the VMD divides the screen area into detection zones, and a change in one or more of these can trigger an alarm. By using zones the VMD can be given a degree of intelligence. For example, if used in an outdoor location, a sudden darkening of the entire scene caused by a cloud passing across the sun can be ignored by the VMD. Similarly, where one zone is equivalent to only a very small area in the field of view, the unit can be programmed such that an activation of just one zone will also be ignored, as in all probability the would-be intruder is more likely to be a bird or other small creature.

The unit is designed to detect changes in the contrast level in each detection zone, and the sensitivity may be adjusted by altering the amount of contrast change required to initiate an alarm.

Analogue VMD units have been available for many years, but their sensitivity was limited, making them prone to false alarms due to sudden changes in lighting level, movement of small animals, etc. Digital VMD units take the analogue signal and immediately pass it into an A/D converter. The reference picture is placed into a frame store so that each subsequent frame can be compared with this image. Digital units are capable of breaking the screen up into many thousands of detection zones, giving them a great deal of sensitivity, and the operator is able to control this sensitivity using two parameters: the amount of grey scale change and the number of zones over which this change takes place. To make the sensitivity adjustment simple for the operator to use, the unit often has some form of sliding scale on the OSD which is calibrated using arbitrary numbers.

The operator is normally able to set the detection area(s) for each camera by using an array of rectangles, circles, etc., which are generated by the OSD. This is illustrated in Figure 9.16. Note that these detection areas are not to be confused with the much smaller detection zones used by the VMD microprocessor for image analysis. Also note that the letters and numbers in the illustration are for the purposes of reference only, and would not appear on an actual screen.

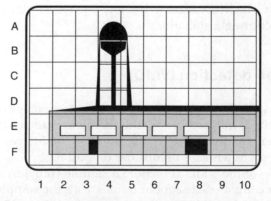

Figure 9.16 *A video motion detector divides the image into a series of zones. The operator (or engineer) decides which zones are to be active*

In Figure 9.16, let us suppose that there is legitimate movement in front of the building, but the operator is only concerned about intruders moving along the flat roof area. The rectangles between points D2 to D10 (and possibly C2 to C10) could be made active, thus causing an alarm to be generated when there is any movement in this area.

The majority of MUX units incorporate VMD as standard, making them an attractive choice of CCTV control equipment in small- to medium-sized systems, especially if telemetry is also included.

VMD can also be found in some cameras that have digital processing, the alarm condition being present on a pair of switch contacts which can be connected to a MUX or other alarm input.

An important point with VMD is that the camera must be reasonably stable. If it is prone to movement in high winds, etc., the movement of the image can result in false alarms. Some VMD processors have in-built compensation for this effect, but nevertheless every system has its limitations, and a secure mount is recommended.

10 Telemetry control

Without some form of remote control, larger CCTV installations would be of limited use. The operator needs not only to be able to adjust the angle, zoom and focus of cameras, but also to perform other 'housekeeping' tasks such as washing and wiping of the front glass on external housings and control of lights. Some of these commands are quite simple. Take, for example, the wash command; it is simply an on/off situation. However, other control commands are more difficult to send. There is little to be gained from simply 'telling' the pan motor to 'pan'. Perhaps if the motor could answer back it might ask, 'Which way? For how long? How fast?'

Operators of fully functional cameras have come to expect a lot from their systems, and to some extent are attempting to emulate the actions of a broadcast camera operator, but from a distance. The number of commands needed to provide such complex control is substantial, and requires sophisticated telemetry to enable adequate communication. A list of the main commands is given in Table 10.1. When the command includes speed, this refers to controllers that have a dynamic (or ballistic) joystick where the further the stick is pressed, the faster the motor moves. Thus, for example, an instruction may not just be 'pan left' but rather 'pan left at this speed'.

Table 10.1 *Major telemetry control commands*

Command	Number of instructions
Camera address (i.e., number)	1
Pan left + speed	2
Pan right + speed	2
Tilt up + speed	2
Tilt down + speed	2
Zoom in	1
Zoom out	1
Focus far	1
Focus near	1
Iris open	1
Iris close	1
Wash on/off	2
Wipe on/off	2
Lights on/off	2
Move to preset position *n*	Dependent on the number of presets

From the table we see that there are no fewer then twenty-one commands for a fully functional camera, plus an additional number for the preset positions. In this chapter we shall look at ways of communicating this information to the various cameras in a system.

Control data transmission

Some of the earliest remote control CCTV systems relied on a hard-wired link between the control console and the PTZ head. However, these systems required a lot of multicore cabling between the control room and each individual camera site and, in some cases, were prone to the effects of voltage drop along the cables.

A much more effective alternative to this method of control was to send digitally encoded PTZ commands along a twisted pair cable in the form of RS 422 or RS 485 (see later in this chapter). The principle is illustrated in Figure 10.1. Each command is encrypted into a data format and sent along a two-wire link to a receiver. The receiver (site driver) contains a decoder chip which interprets the commands and operates the appropriate relay(s) via the relay driver chip.

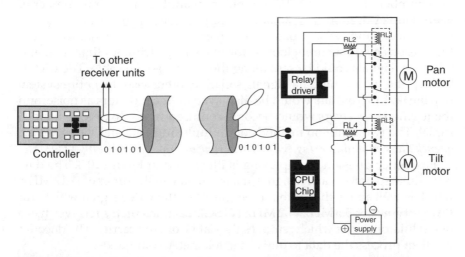

Figure 10.1 *Control of motors using a separate data link such as RS 485 over a twisted pair*

There are two ways that the data links may be connected to the receiver units. The first is like that illustrated in Figure 10.1, where the controller uses individual outputs to each receiver. The other way is to place the control data onto all of the lines simultaneously, enabling receivers to be connected individually to the controller and/or daisy chained. This is illustrated in Figure 10.2.

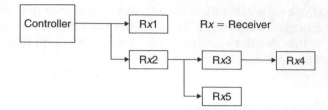

Figure 10.2 *Simultaneous data transmission*

Where a daisy-chain design is used, the encoder includes an address in the command. For example, if the operator selects camera 5 to pan left, then the encrypted data is effectively saying, 'camera 5, pan left'. Note that the first part of the command contains an address – 'camera 5'. Thus, although the signal is picked up by all of the receivers, only the one which has been assigned as camera 5 responds. The receivers are assigned their addresses during commissioning. Note that a controller designed for individual output transmission cannot be connected daisy-chain fashion because it will not be sending any address data.

Twisted pair telemetry transmission is very effective and was employed by a number of leading CCTV equipment manufacturers for a number of years. However, from an installation point of view, the requirement to run both co-axial and twisted pair cables to every camera was not ideal. For this reason, manufacturers looked for ways in which both the video and telemetry signals could be sent along the same co-axial cable. The trick is to keep the video and data signals separate, otherwise the entire system will break down as data would be displayed on the monitors as noise, and the telemetry receivers would become confused trying to decode a video signal. The two common methods of multiplexing data and video are *frequency division multiplexing* (FDM) and *time division multiplexing* (TDM).

Frequency division multiplexing is illustrated in Figure 10.3. The data is modulated onto a carrier signal at a frequency in the order of 8–12 MHz, which is well above that of the upper limits of the video signal which, for PAL, is typically 5.5 MHz (4.2 MHz NTSC). Each telemetry receiver has a demodulator circuit which removes the data from the carrier. The decoder can then process the data to derive the telemetry commands.

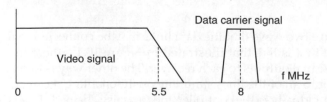

Figure 10.3 *Signal spectrum for frequency division multiplexing*

Although popular for a number of years, the majority of CCTV telemetry equipment today operates using time division multiplexing, which is a take-off of the Teletext system used in the UK broadcast television service. Here data is transmitted during the field flyback blanking interval, during which time the electron beam in the CRT is cut off whilst the scanning circuits adjust themselves to begin scanning the following field from the top of the screen. When viewed on an oscilloscope, the signal appears similar to that shown in Figure 10.4. As with FDM, the telemetry receivers each have a decoder circuit which picks out the data transmission and deciphers the commands.

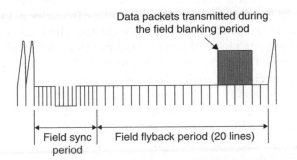

Figure 10.4 *Time division multiplexing. The data is sent at a different time to the video information*

The TDM signal is at a frequency somewhere in the region of 4.5 MHz, and thus we see that in the case of both TDM and FDM, it is essential that the bandwidth of the system is up to specification. Any faults in the cabling or termination that would introduce a high-frequency filtering effect will not only remove the high-resolution components of the picture, but may also filter out the telemetry data. Such faults can be intermittent, or may only affect certain camera locations.

Pan/tilt (P/T) control

The pan and tilt unit comprises two motors and a number of gears to convert the motor speed into torque. The motors may be 24 Vd.c., 24 Va.c. or 230 Va.c. Whilst a.c. motors are generally more efficient and often produce greater torque than their equivalent-size d.c. counterparts, speed control of an a.c. motor is somewhat more complex than for d.c. motors. Thus, single-speed 230 Va.c. motors are ideal where it is anticipated that high winds are likely to exert a heavy load on the camera assembly, and a high torque drive mechanism is required to overcome this. However, where dynamic joystick control is to be incorporated in the system, d.c. P/T units will allow multi-speed (where the speed alters in incremental steps) or variable-speed operation. It is usual for the motor to be coupled to the gears via clutch assemblies which will slip in the event of the mechanism jamming, thus protecting the motors from stalling and burning out.

For simplicity the circuit in Figure 10.1 shows d.c. motors. Where a.c. motors are connected to the driver circuit the arrangement is similar; however, simply reversing the polarity of the a.c. supply to the motor will not cause it to reverse. To achieve this action, the relay changeover contacts must reverse the connections to the motor field windings. A.c. motor theory is not something that the CCTV engineer needs to become too involved with; however, it is important when installing or replacing P/T units to ensure that they are compatible with the drive voltage on the site driver. Applying 230 V across a 24 Vd.c. motor is likely to have disastrous results! Some site drivers have the facility to select the type of drive voltage (see later in this chapter).

A P/T unit contains at least four limit switches. These are a form of cutout which are adjusted to set the maximum points of deflection in both the horizontal and vertical directions. Without these the motors could simply drive the unit in circles, causing the interconnecting cables between the housing and the receiver to become twisted around the mounting until they break, or until the housing becomes jammed. During installation the engineer adjusts the limit switch positions to suit the particular location, but in most cases the unit is designed such that it will not pan through more then 355° before stopping.

Limit switches are satisfactory if the camera is only going to be moved occasionally; however, with permanently manned systems the need became evident for a facility whereby a number of predetermined positions could be set for each fully functional camera location. For these presets to be effective it is not only necessary to fix the pan and tilt positions; the zoom and focus settings for that position must also be fixed. Clearly mechanical limit switches are not able to offer this amount of control, and some other means had to be devised.

The most common solution is to place variable resistors inside the pan/tilt unit and the zoom lens assembly (this was discussed in Chapter 4, Figure 4.19). As the motors move each of the pan, tilt, zoom and focus mechanisms, the associated potentiometer rotates, producing a d.c. feedback voltage which can be monitored by the control circuit. During programming, the operator moves the P/T unit to the desired position, adjusts the zoom and focus, and enters these into a memory store, allocating a preset number. At this instant the control unit measures the d.c. feedback from each potentiometer, passes this to an A/D converter, and stores the digital values in memory. During operation, whenever the preset number is selected, the control unit checks the position of the potentiometers, and moves the four motors until the feedback voltage from each potentiometer is equal to the value stored in memory.

With this control method the presets and limit positions are not set mechanically, but in the PTZ controller. Thus there is, in theory, no limit to the number of possible preset positions. Depending on the controller, as few as five, and as many as one hundred, preset positions can be programmed.

A large number of positions is not a lot of use to a CCTV operator because it is not feasible to expect him/her to remember every one. Consider a ten-camera system where each camera has one hundred preset positions programmed: who could remember one thousand preset position numbers? However, many control systems have an *automated patrol* facility whereby, in the absence of any activity or perhaps at times when the system is unmanned, the operator can leave the cameras moving through a set pattern of positions. It should be pointed out that where this facility is used extensively, the wear and tear on the P/T unit gears and clutches can take its toll.

Another use of multiple presets is in conjunction with alarm signals (see Chapter 9). Upon receipt of a particular alarm signal input, the controller can often be programmed to move one or more cameras to a given position and prioritize these video signals in the recording process.

Receiver unit

A typical telemetry receiver unit is shown in Figure 10.5. In this illustration we can identify many of the features we have discussed so far in this chapter.

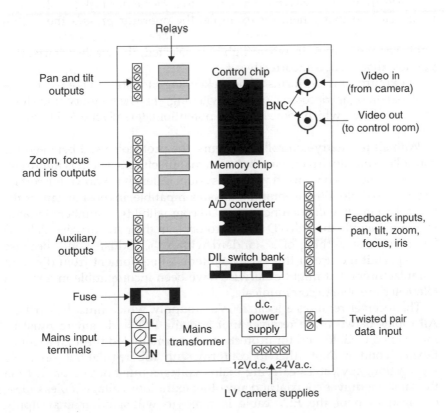

Figure 10.5 *Typical receiver unit*

Where the data is multiplexed with the video signal on one co-axial cable it is necessary for the camera signal to loop through the receiver in order that the data can be extracted (bear in mind that the video signal is passing from the camera to the control room, whereas the data is going the other way). Some receivers may have the option of a twisted pair telemetry input, and this brings us to the next point, the DIL switches. These are used to customize the receiver to the system and location, providing such options as co-axial or twisted pair data input, 4-wire or 3-wire lens wiring, 5 V or 12 V lens operation, PT motor drive voltage, auxiliary output functions, etc. The correct setting of these switches during installation is of utmost importance as the unit will not function correctly (if at all) if any of these are in the wrong position for the type of installation. Manufacturers' instructions must be carefully adhered to.

For the driver in Figure 10.5, the pan and tilt outputs are effectively those we looked at in Figure 10.1; however, the auxiliary outputs are somewhat different. Because these may be used for switching either mains-voltage or low-voltage devices, the terminals are simply taken to a set of clean relay contacts on the PCB. The installer can therefore connect the required power for the device through these switch contacts. The DIL switches are used to assign each output to a particular task (wash, wipe, lights, etc.) so that when, for example, the operator presses the button labelled 'lights' on the keyboard, the auxiliary output assigned to the lights changes states. If, when 'lights' is selected, the washer starts, it is possible that a DIL is incorrectly set.

The control chip performs all housekeeping functions in the receiver as well as data signal decoding, output switching, etc. The memory chip stores the preset levels which come from the potentiometers via the A/D converter chip.

With all telemetry-controlled systems, the protocols used between the controller and site drivers vary between manufacturers, and it is important that equipment is not mixed within a system unless it is known for certain to be compatible. Where equipment is incompatible, in most instances the telemetry simply will not function. To aid compatibility, a number of manufacturers include the Pelco D set of protocols in their site drivers. Pelco D has become something of a standard within the industry and, because Pelco permit the use of this protocol, items of equipment from different manufacturers that might otherwise have been incompatible in terms of telemetry are able to communicate.

The receiver requires a local 230 Va.c. supply, which must be earthed. All cables between the receiver unit and the P/T and camera must be made to be flexible, and for protection are normally fed through a Copex flexible conduit. Note that to conform with IEE regulations on cable segregation, any cables carrying mains voltages must not be passed via the same conduit as co-axial and any other extra-low-voltage cables unless the insulation of the ELV cable is rated to withstand mains supply voltages.

Dome systems

Pre-assembled domes containing a P/T unit, site driver and power supply have largely overtaken the traditional pan/tilt unit. In some cases a camera is included as a part of the assembly, but this can be somewhat limiting as the installer then has no choice of camera specification. There are domes available where a camera and fully functional lens of the installer's choice can be fitted. Internal and external versions are available.

There are perhaps a number of reasons for the popularity of domes. Very often they are preferred by the customer because they are overt, and yet covert. That is, people can clearly see that the premises are monitored by CCTV, but they never actually know when they themselves are being observed. Aesthetically, domes often look less out of place than traditional PT and site driver assemblies. Installers have less work because they are ready assembled. But perhaps one feature above all of these is the availability of high-speed domes, which are capable of 360° continuous rotation.

The high-speed function enables the PT unit to move the camera at rates in the order of 300° per second, with a corresponding response from the zoom and focus servos. The inclusion of multiple presets offers rapid response to operator or alarm-activated requests. The speed control is made possible by the inclusion of sophisticated servos on the site drivers controlling either d.c. stepper motors or, in some instances, a.c. synchronous motors using a variable high-frequency supply voltage. (Note: altering the frequency of the a.c. supply to a synchronous motor changes its speed.)

The 360° rotation is made possible by the use of slip rings to make all of the electrical connections, including the video signal, between the camera and the site driver. This principle is illustrated in Figure 10.6.

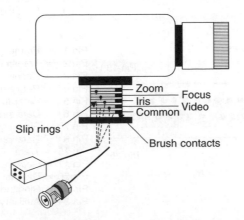

Figure 10.6 *Principle of using slip rings to couple a rotating camera to fixed cables*

Data communications

In Chapter 11 we will take a detailed look at network communications appertaining to both CCTV signal transmission and remote control of CCTV devices. However, TCP/IP over a LAN/WAN is not the only method for telemetry control of CCTV equipment, and a number of common standard interfaces are still employed, including *RS 232, RS 422* and *RS 485* (RS = recommended standard) to perform functions like telemetry control and equipment connection. To enable equipment of different industries and manufacture to communicate, some means of standardization of signal level, polarity and frequency is necessary. Also, the configuration of connectors between equipment must be the same. Such standardization is found in the recommended standards.

Perhaps one of the oldest communication protocols used in the computing industry is the RS 232, which was introduced in 1962. This is a very common interface and has been used extensively for many years to link computer-based equipment. In most cases the connector type known as the *D connector* is used; the pin configurations for both the DB9 (9-pin) and the DB25 (25-pin) are shown in Figure 10.7.

The voltage levels for RS 232 are much higher than for TTL logic circuits. In theory, logic 0 is between +3 V and +25 V and logic 1 is between −3 V and −25 V. However, in practice, the majority of RS 232 drivers operate

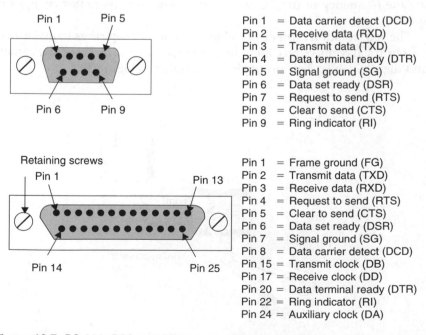

Figure 10.7 *RS 232, DB9 and DB25 type connectors. (Male connectors, viewed from front)*

between ±12 V, and there are some drivers that enable RS 232 data to swing between just 0–5 V, although the cable length may be limited.

The main limiting factor with regard to RS 232 is the relatively short maximum cable length, which is specified as 15 m (50 ft). The main reason for this short length is the amount of capacitance in the cable which, as we saw in Chapter 2 (Figure 2.2), integrates the square pulses corrupting the data signal. Another limiting factor is the low baud rate (the speed at which data is transmitted, expressed in bits per second or bps). The maximum specified speed for RS 232 across a 15 m cable length is 20 kbps, although rates of up to 115 kbs may be achieved over shorter distances.

Referring to Figure 10.7:

- *Frame ground* is a protective earth which protects the sensitive microchips from damage which would occur if static were allowed to build up around equipment cases. It also serves as a screen, preventing RFI and EMI from corrupting the data.
- The *transmit data* and *receive data* connections are the main data links from between the two items of equipment at either end of the bus.
- *Request to send* is used for modem control. It switches on the carrier when the data terminal equipment (DTE) is ready to send data.
- *Clear to send*: in response to a logic 0 on RTS, the modem takes CTS to logic 0 as soon as it is ready to transmit. In other words, it is telling the DTE that it is ready. *Data set ready*: when the modem is active, it takes this pin to logic 0 to let the DTE know that it is connected to a 'live' modem.
- *Signal ground* is a common negative to which all signals are referred.
- *Data terminal ready*: this is the complement to the data set ready action. The clock signals maintain correct timing/synchronization between devices for the purposes of transmitting and receiving data.

The activity between RTS and CTS is known as *handshaking*, the term describing the co-operation between devices that are exchanging data. This protocol is necessary because it is not possible for equipment to send and receive at the same time. A typical handshake sequence would be: DTE asserts RTS – slight delay whilst the modem starts up – modem responds with a reply on CTS – DTE begins transmission of data.

The limitations of RS 232 are overcome in standards such as RS 422 and RS 485, both of which make use of low-impedance differential signals passing along a twisted pair cable. Much longer transmission distances are possible (in the order of km) and, because any external noise is induced equally in both conductors, the noise is cancelled as a result of the action of the balanced cable plus the balanced inputs at the receiving equipment. Special line amplifiers/drivers are required to launch and receive the differential signals. The principle is illustrated in Figure 10.8, whilst the principles of twisted pair transmission were considered in Chapter 2.

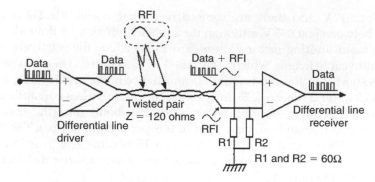

Figure 10.8 *Illustration of RS 422/485 transmission. RFI is cancelled both by the action of the balanced cable, and the differential inputs.*
R1/R2 form a 120 ohm termination resistance across the line; note the reference to earth (see text)

RS 422/485 are described as two wire (single twisted pair) systems. Whilst this is true, the line drivers also rely on a ground connection to derive the signal voltage and, bearing in mind what was discussed in Chapter 2 regarding ground potential differences (Figure 2.11), problems may arise where the potential at each end of this ground connection is different. In order to provide a reliable ground connection between devices, many manufacturers specify double twisted pair cable, where the second pair are connected in parallel to the ground terminals. The doubling up of conductors ensures a low electrical resistance, and the ground connection is much more reliable than an electrical earth because there is no possibility of a ground loop current forming.

The RS 422 standard defines a maximum data rate of 10 Mbps at 15 m (50 ft) maximum cable length, and 100 kbps at a maximum cable length of 1.2 km (4000 ft). The signal level is between 0 and 5 V. One feature that makes RS 422 popular for many applications is its ability to interface easily with RS 232. By connecting a simple line driver to a PC COM port, the RS 232 signal can effectively be transmitted over a distance of some 1 km by converting it to RS 422, and then using a second line driver at the receiving end to restore the signal to RS 232 protocol once more.

The RS 422 bus can include up to five transmitters and up to ten receivers, but only one transmitter may connect to the line at any one time.

Introduced in 1983, RS 485 is able to support up to 32 devices on one twisted pair, although modified versions are able to support up to 256 devices. Devices are connected in parallel across the twisted pair in a daisy chain, and each device may be configured to be both a transmitter and receiver although, like RS 422, only one device may transmit at any one time. The line drivers are tri-state devices, which means that their (transmitter) output may be at logic 0 or logic 1, or disconnected from the line.

In a CCTV application, the first device may be a telemetry controller, and the other 31 devices site drivers. Where the system is to be administered by a PC, the RS 485 network may be connected to the PC via either an RS 232 (COM) port or USB port in a similar manner to that described above for RS 422. An illustration of a PC administered system is given in Figure 10.9.

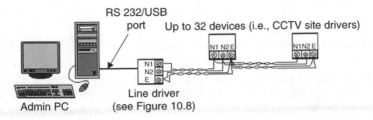

Figure 10.9 *The line driver converts the RS 232 or USB output to RS 485. Devices are then daisy chained across one pair, the second pair serving as a ground wire*

In order to prevent signal reflections in the RS 485 data bus, the two ends of the bus should be terminated at the correct impedance. The termination impedance is typically between 100 and 120 Ω, depending on the type of cable employed, and the design of the differential line drivers (Figure 10.8).

Both RS 422 and RS 485 specifications exclude any actual transmission protocol, meaning that the manufacturer is free to use or develop a protocol that best suits their application. This may explain why one CCTV site driver that employs RS 485 is not necessarily compatible with other site drivers that also use RS 485.

11 CCTV over networks

The combination of high-speed data transmission capability and effective video compression techniques has resulted in an inevitable shift towards using IT network technology to transmit CCTV video images, both locally and over long distances. Consequently, the CCTV engineer, who in many cases began his working life as a time-served electrician or security/fire alarm systems installer, now needs to have reasonable knowledge and competence in both computer-operating systems and networking. It certainly is not the intention to turn security systems engineers into fully fledged network engineers by the end of this chapter, but the reader will have an understanding of network topology, how data is managed across networks, the limitations of any network, and how all of this applies in a practical way when connecting IP cameras and network video recorders (NVRs) and the like onto a network.

Network topology

A topology is a structured wiring diagram, or map, of a network. In computer networks there are five topologies in use (although some are far more common than others): bus, star, ring, mesh and wireless. These are illustrated in Figure 11.1.

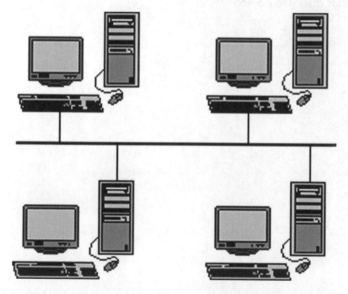

Figure 11.1a *Bus topology. All devices are connected to a single continuous cable*

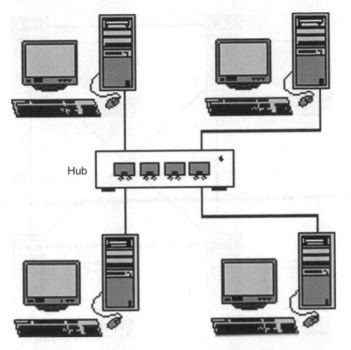

Figure 11.1b *Star topology. Often used in conjunction with bus topology*

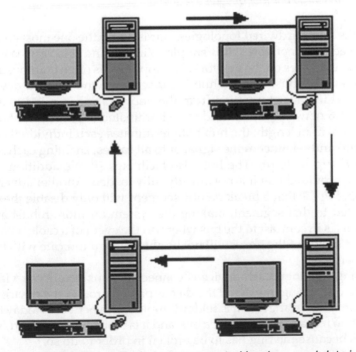

Figure 11.1c *Ring topology. Devices are connected in a loop, and data is passed in one direction between devices*

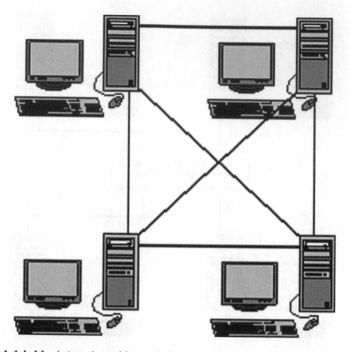

Figure 11.1d *Mesh topology. Very reliable, but difficult to install and modify owing to the large number of cables required*

Of the four hardwired topologies, star is by far the one most commonly employed in IT networks. It is simple to install and it is very easy to add or remove devices on the network. Star topology is often employed in conjunction with a bus, where a hub is connected to the bus and devices are connected in a star formation from the hub. Bearing in mind that CAT 5 and CAT 6 network cables need to be terminated at $100\,\Omega$ and should not exceed 100 m in length, the hub both terminates each individual cable (or segment) and relaunches the signal onto all cables, enabling each segment to stretch up to 100 m. The hub also facilitates simple addition of other devices, provided that it is not already fully loaded. Another advantage of star topology is that a break on one segment will only disable the devices connected to that segment, making the system far more robust and reliable. This is in contrast to the bus where, in the event of a cable break, devices on one side of the break will not be able to communicate with those on the other.

In ring topology, each device is connected to its neighbour via a loop in/loop out arrangement and the data is passed around the devices in one direction. This arrangement is seldom employed in networks today because it offers no real advantages over a bus, and it is more difficult to add/remove devices because the ring has to be broken in order to do so.

Mesh is a theoretical concept and would, in practice, prove far too expensive to install, and it would be very difficult to add/remove devices.

The illustration in Figure 11.1d only shows four devices, but imagine the problems in connecting fifty or a hundred devices in this manner! In practice, mesh topology is used to a limited extent on the Internet in order to provide reliable connection.

RF wireless networks, sometimes referred to as ad hoc networks, are popular because of their versatility; all that is required to establish a connection is the presence of a signal between the device you are using and the network you are connecting to. Wireless networks have a number of very obvious advantages including portability of devices, simple addition/removal of devices and limited installation costs. Nevertheless, because of bandwidth limitations, wireless connections are generally slower than hardwired, and in some cases the signal strength may vary over time, resulting in intermittent loss of connection.

Networks are divided into two types: *local area networks* (LANs) and *wide area networks* (WANs). It is sometimes difficult to determine exactly where one ends and another begins, and there are differing interpretations of this terminology. In simple terms a LAN can be defined as a cluster of PCs (or other network devices) that are able to pick up an Ethernet broadcast and, as we shall see later in this chapter, broadcasts are blocked by routers. So a LAN is generally confined to one side of a router. By contrast a WAN spans across routers and, although it does not necessarily have to cover a wide geographical area, in many cases it connects between sites many miles apart. It should be pointed out that a WAN does not have to include the World Wide Web (or Internet). It may be a private network comprised entirely of private lines connecting the sites. Such a WAN is often referred to as an *Intranet*.

Network hardware

A typical WAN comprised of two LANs that employ both bus and star topologies is shown in Figure 11.2. In this topology, the *backbone* connection is the main bus onto which all network devices connect. Each individual connection to the backbone is known as a *segment*. The backbone usually has a much broader bandwidth than the segments as it has to carry all of the traffic on the network.

Any device that is connected to a network is known as a *node*. Most devices connect to the network via a *network interface card* (NIC), and most of these are addressed on the network using an IP address (see later in this chapter). A network device that uses an IP address is referred to as a *host*. The NIC, also known as the Ethernet card, performs the functions of organizing the data into frames, managing the transfer of these frames between hosts, and error correction. The NIC also plays an important role in determining the speed of the network. In Chapter 2 we looked at possible data speeds for CAT 5 and CAT 6 structured cable networks. However, it is not only the cable that determines the data rate; the NICs must also be capable

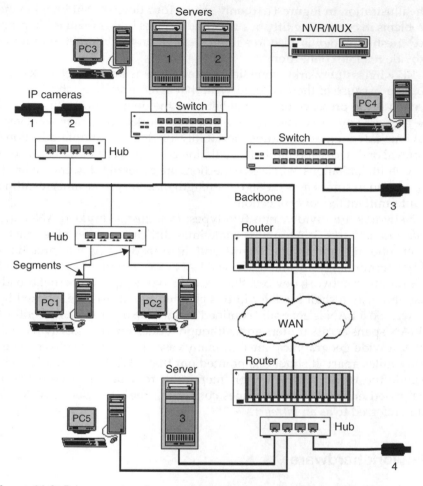

Figure 11.2 *Primary network components, including CCTV devices. This example shows a WAN comprised of two LANs*

of processing data at the required rates. For example, connecting a device that has a CAT 5 NIC to a CAT 6 network will not produce any improvement in connection speed as the system will always be forced to operate at the speed of the slowest device.

Every NIC is given a unique *media access control* (MAC) address, also known as the *hardware address*, or the *Ethernet address*. This takes the form of a twelve-bit hexadecimal code, separated by hyphens. For example,

    00-20-4A-32-F2-3C

The Institute of Electrical and Electronic Engineers (IEEE) has been assigned the task of administering MAC addresses throughout the world, and any manufacturer of NIC or other network devices will purchase a unique address comprised of the first six bits of the twelve-bit code; i.e., 00-20-4A in the example above. By making the second six bits of the MAC address unique, the manufacturer is able to assign a unique MAC address to every device they produce. As we shall see later in this chapter, the MAC address plays a very important role on the network.

In a device such as a PC, the NIC is normally integral to the motherboard, although it may be fitted to one of the PCI slots. In a device like a network camera (or any other network security device such as an alarm collector point or access control door controller), the NIC is often incorporated within the on-board RJ 45 socket.

A hub is perhaps the simplest of all network devices because it has no intelligence and makes no decisions. It simply provides a star connection for (typically) four or eight devices, with a 100 Ω termination for each connection. Most modern hubs are active devices, which means that they incorporate separate amplifiers for each input/output connection to ensure reliable data transmission without attenuation. The amplifier action also serves the purpose of doubling the distance between one host and another. In Figure 11.2, the maximum distance between the NVR and IP cameras 1 and 2 would be limited to 100 m. The inclusion of the hub effectively doubles this to 200 m. There is a limit as to how many hubs may be used as repeaters on one line to extend the cable length, because eventually the S/N ratio will become so poor that connection between NICs will be lost.

In terms of data transmission, a hub does not reduce network traffic because it simply passes all traffic to all hosts. For example, in Figure 11.2, the images from IP cameras 1 and 2 may be intended only for the NVR, but the data packets from these cameras will pass through both hubs and will be present on the segments connecting the two PCs. This will slow down that part of the network because the PCs will have to wait their turn to gain access in order to transmit their information.

A much more effective way to create network segments is to use a *switch*. Although at first glance it appears like a hub, a switch is intelligent enough to recognize the intended destination for any data packets and will only place those packets on the corresponding output. Thus, a switch creates a dedicated path between any two hosts for the duration of their communication. This action greatly improves network performance because it means that, for much of the time, more than one pair of hosts can communicate at the same time. Taking the example in Figure 11.2, if PCs 1 and 2 were connected via a switch rather than a hub, communication between these two hosts would not place any traffic on the backbone, and other hosts connected on the backbone would be able to communicate at the same time.

In some instances, the use of a hub may be preferable to a switch because the processing action in the switch does introduce an element of delay, whereas a hub passes the data much more quickly.

The way that a switch identifies the destination of data packets can vary, depending on the type of switch. Some switches look at the MAC address (layer 2 switches) whilst others look at the network address for the data packets. These are known as layer 3 switches, although in practice they are really routers. The type of switch, therefore, can affect the way that the network functions and how certain devices (such as IP cameras) connect to their hosts.

Similar to the switch is the *bridge* which, like the switch, intelligently connects segments together. The main difference is that, whereas a switch connects multiple segments (like a hub), the bridge only connects two segments. It is used to keep traffic on the two segments separate except when it is necessary to pass it over. The principle is illustrated in Figure 11.3. There are two types of bridge, those that are able to connect dissimilar networks (i.e., ring to bus) and those that can only connect similar networks.

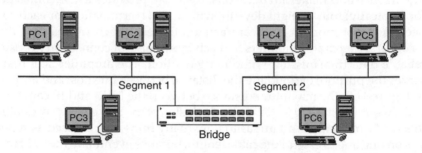

Figure 11.3 *The bridge only passes traffic across segments when it is necessary to do so, for example, when PC1 is communicating with PC4. This action improves the network speed because, at times when the segments are isolated, hosts on Segment 1 can communicate at the same time as hosts on Segment 2*

A *router* is used to send data from one segment to another or from one network to another and, like a switch, it reduces network traffic by only forwarding data when necessary. However, unlike either a switch or a bridge, a router is an intelligent device that is on a par with a PC. Routers constantly interrogate the network to find out what other networks and routers are out there and then store this information in their *routing tables*. When a host attempts to send data to a host on another network, the router will use its routing table to determine the shortest path through which to send the data, bearing in mind that the path may include a number of routers.

Another significant difference between a router and a switch is that a router will not pass what are known as broadcast data transmissions (these will be discussed later in this chapter). This action can sometimes affect the setting up of network devices such as IP cameras because the connection protocol used for the initial setup may be a broadcast, meaning that the router prevents the engineer from connecting to the device in order to perform set-up. This is why it is sometimes necessary to configure the IP camera on the network local to the administration PC before deploying it on site.

The engineer may be requested to enter the *gateway IP address* when setting up IP devices. A gateway device is the one through which all data must be routed in order for it to be passed on to another network (either local or wide area). Consequently, a gateway is normally a router, but it does not have to be. In some cases it may be a PC or a server that is acting as a gateway. Note that, where a device such as an IP camera only communicates with hosts on the same subnet, it is not necessary to enter a gateway IP address.

A network *server* often looks like any other PC tower, although it may be a little larger. In truth, there is not a lot of difference between the system architecture of a PC and a server. A server usually has a much higher specification than a normal PC in terms of processing power and is designed to operate 24/7 over a period of years. However, what really makes a server perform as such is the operating system that is installed, for example Microsoft 2003 Server, Redhat Linux, or Sun Solaris.

The function of servers on a network is to provide services to the client PCs and other devices connected to the network. Such services include DHCP and DNS (discussed later in this chapter), mail services, web connection, application software, print services, file management, etc.

Network communications

The network communications protocol most commonly encountered is *TCP/IP*, or transmission control protocol/Internet protocol although, as we shall see later, there are other protocols (or languages if we want to view it from a human perspective) in use. The IP address most commonly associated with TCP/IP is IP Version 4 (IPv4), which has been with us for a considerable number of years now, having been devised in the 1970s.

A protocol is simply a set of rules, so TCP/IP are two sets of rules that are applied to the function of transmitting data over a LAN/WAN. The transmission control protocol is responsible for ensuring that data is transmitted across a network accurately, that an acknowledgement is sent by the recipient host, that data is re-sent if no acknowledgement is received, that data is checked for errors at the receiving end, and that the data packets are re-assembled in the correct order. It achieves this by first breaking the data into packets and then adding a TCP header to each packet, which is a bit like placing your letter into a stamped and addressed envelope before posting it. These data packets with their TCP headers are known as *datagrams*, and each datagram must have a minimum length of 576 bytes and a maximum length of 65 536 bytes.

The TCP processes are perhaps best understood by examining the packet header, which is shown in Figure 11.4. The source and destination port information ensures that data is sent between the intended processes in each host. The sequence number contains the necessary data to ensure that the packets are in the correct order for reconstruction. The acknowledgement number is used by the receiving host to acknowledge to the source host that the packets have been received. If no such acknowledgement is

Source port
Destination port
Sequence number
Acknowledgement number
Offset
Reserved
Flags
Window
Checksum
Urgent pointer
Options
Padding
User data
0
16
32
bits

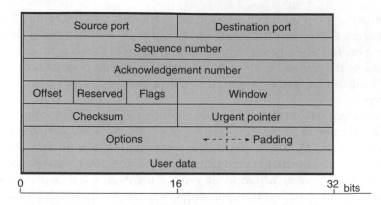

Figure 11.4a *A datagram with its TCP header*

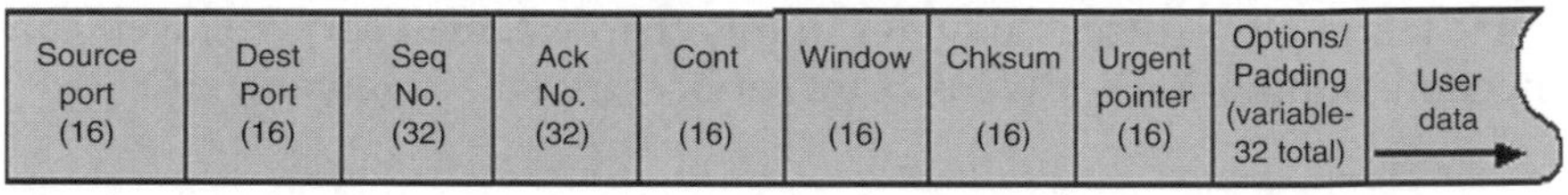

Source port (16)
Dest Port (16)
Seq No. (32)
Ack No. (32)
Cont (16)
Window (16)
Chksum (16)
Urgent pointer (16)
Options/ Padding (variable- 32 total)
User data

Figure 11.4b *Transmission sequence for a datagram. The numbers in brackets indicate the number of bits*

received after a certain period, the packet will be re-transmitted. The checksum data is used to check received data for errors. When errors are detected in a packet, that packet is discarded, the acknowledgement number is not incremented and the sender will re-transmit the data. The other data in the packet header, although important, is beyond the scope of this book.

Once TCP has created the datagrams, they are sent to the Internet protocol for final packaging before transmission. IP has the task of adding the destination address, which will be used by all routers and gateways to determine the optimum transmission path in terms of length and speed. IP is connectionless, meaning that there is no handshaking or acknowledgement of receipt, which is why it is used in conjunction with TCP in order to guarantee reliable data transmission. The IP header also contains information relating to the IP version, the length of the datagram, its own checksum, a hop count, and other information. The *hop count* is normally set at 32, and this determines the maximum number of network routers that the datagram may pass before it is discarded as undeliverable. If there were no hop count, undelivered data would simply continue to pass around the Internet indefinitely like some sort of refugee, and by now the web would be unusable because it would be congested with such data. The complete datagram containing data plus TCP and IP headers is shown in Figure 11.5.

An IPv4 address is made up from four 8-bit words known as *octets*, each octet being separated by a dot. An 8-bit binary word can have up to 256

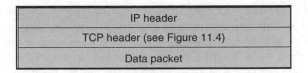

| IP header |
| TCP header (see Figure 11.4) |
| Data packet |

Figure 11.5 *A complete datagram containing its IP and TCP headers*

possible permutations and therefore, rather than denote the octets in binary, which would be difficult for us to remember, they are denoted in their denary equivalent values. For example, the binary octet 11 000 000 would equal 192 when converted to decimal, so we show it as 192 in the IP address. Thus, all octets in an IP address will have a value between 000 and 255, and the complete IPv4 range of possible addresses will be between 000.000.000.000 and 255.255.255.255.

IPv4 classes

In the original conception, IPv4 addresses were broken down into classes. The classes that we are interested in are classes A, B, C and D. The primary difference between classes is the way that the network and host addresses are denoted, which is illustrated in Figure 11.6 for classes A to C. Here we can see that some octets contain the network address, whilst the remainder contain the host address. It is often said that the network address can

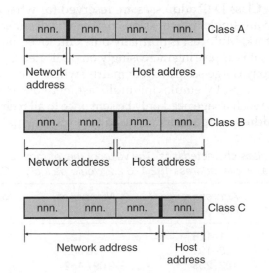

Figure 11.6 *Each IP class uses different octets to denote the network and host addresses*

be likened to a street name and the host address can be likened to the house address. This is illustrated in Figure 11.7, where two separate networks are connected via a router. PC1 and PC4 both have the same host address (.11) but their IP addresses are different because of their differing network addresses: PC1 has an address of 194.176.2.11, whilst PC4 has an address 194.176.3.11.

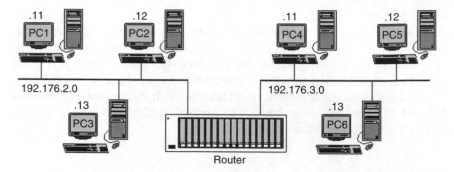

Figure 11.7 *This illustration depicts two networks that have different addresses; however, the hosts have the same address range. Note that the network address uses a zero to denote the host octets, which is why all network addresses end with a zero. The IP address of PC1 is 192.176.2.11, whilst the IP address of PC4 is 192.176.3.11*

The IP class was originally denoted by the value of the leftmost octet. These values are shown in Table 11.1, where it will be noticed that value 127 is not included. This is because it is reserved as a test address, known as a *loopback address*. Class D IP addresses are reserved for what are known as *multicast*, which are single transmissions that can be destined to groups of hosts on a network. Multicast is normally only employed on research networks; however, it is appearing increasingly on dedicated CCTV networks where images from a large-scale MUX or matrix switcher need to be accessed by multiple operators. By employing multicast, network traffic is greatly reduced because each image need only be sent once to all recipients. Finally, the class E IP address range is mainly used for experimental purposes.

Table 11.1 *IPv4 class characteristics. The value of the leftmost octet denotes the IP class. For example, address 192.176.2.2 would be a class C IP address*

Address class	Address range (first octet)	Number of networks	Number of hosts
A	1–126	127	16 777 213
B	128–191	16 385	65 533
C	192–223	2 097 152	253
D	224–247	–	–
E	248–255	–	–

Table 11.1 also shows the maximum number of networks and hosts that are available for each IP class. In many respects class C, which is the most common, is the most practical because it permits a lot of networks, but the number of hosts is limited to 253 (in theory this value would be $2^8 = 256$, but the maximum number of available host addresses is always reduced by three, because three addresses are reserved for special functions on the network). Class B may appear more useful because it permits so many more hosts on each network, yet still provides more networks than any institution would require. In practice, to place 65 533 hosts on a single network would result in so many data collisions and subsequent re-sends that the network would prove to be unworkable. This is because Ethernet operates on a principle known as *carrier sense multiple access/collision detection* (CSMA/CD), which in simple terms means that the system allows data collisions to occur and, when they do, the datagrams are simply re-sent. Before sending a datagram, the NIC checks to see if the network is clear and, if it is, the datagram is transmitted. Should another host transmit at the same time the datagrams will collide, and both NICs will back off and wait for a period that is determined by a random timer on each card before attempting to re-send. Therefore one card will re-transmit before the other.

CSMA/CD works well as long as we do not overload the network with too many (busy) hosts. With a very high rate of data traffic, the number of collisions will be very high, which will in turn result in a lot of re-sends, which results in an even greater amount of traffic, resulting in more collisions, and so on.

Reserved addresses

Within the class A, B and C ranges there are ranges of IP addresses that are *reserved*. These addresses are never used on public networks (i.e., the Internet) but are intended for private networks. For class A the reserved address range is between 10.0.0.0 and 10.255.255.255, for class B it is between 172.16.0.0 and 172.31.255.255, and for class C it is between 192.168.0.0 and 192.168.255.255. Another address range that is reserved and will therefore never appear on the Internet is 169.254.0.1 to 169.254.255.254. These are autoconfiguration addresses that a network device will assign to itself in the event of it being unable to obtain an address from a DHCP server (see later in this chapter).

It is important to recognize that it is only essential to adhere to the IPv4 class rules for public networks and that, for a private network, any IPv4 addresses may be employed. At first glance this may appear to be unworkable because we would end up with numerous hosts from different networks communicating over the Internet, all having the same IP address – a bit like having twenty people called John in the room and someone calling in through the door, 'I have a message for John'. However, a private network will never connect directly to the public network but, rather, it must do so through some form of interface. The most common interface is *network address translation* (NAT), where a network device (usually a router or a proxy server) takes datagrams on a private network that are intended for the Internet and removes the local IP (source) address, replacing it with an IPv4 class C or, perhaps, IPv6 address. The principle is illustrated in Figure 11.8. Without NAT we would have run out of IPv4 addresses a long time ago whereas, by using NAT, we are able to use duplicate IP addresses across private networks without conflict.

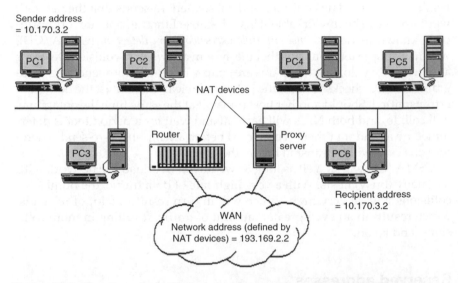

Figure 11.8 *Function of a NAT. By coincidence, PCs 1 and 6 have the same IP address. However, because they belong to different networks, no conflict occurs because the NAT devices change the IP address as the data passes to/from the WAN*

Another advantage of employing NAT is that the proxy server or router makes the original host device invisible to both the Internet and the recipient. This increases the network security although, sadly, in this modern world of hackers and virus script writers, a NAT device alone is woefully inadequate to keep intruders out, and other means of security must be applied such as firewalls, anti-virus software, regular Windows updates, etc.

Subnetting

We have already come to see that we may use IPv4 addresses of any class for private networks; however, we can (and often do) take this a step further. Another data string known as a *subnet mask* is included in the datagram. This data, when overlaid on top of the IP address, is used to identify the network and host parts of the address, instead of using the value of the first octet. So, by using a subnet mask, we can make any IP address fit into any of the classes A to C. For example, by adding a suitable subnet mask to the class A reserved network address 10.172.2.0, we may identify the first three octets as network address and the fourth octet as host address, making it a class C IP address.

One advantage of subnetting is that we are able to create a very large number of new network addresses for private use or, alternatively, we can increase the number of network address by using the subnet mask to move the dividing line between network and host (see Figure 11.6) to the right by one or two places.

With subnetting it is possible to independently route networks. This makes much more efficient use of the available bandwidth by reducing network traffic, reducing the size of the routing tables, providing a simple way of isolating one network from another, and enhancing the ability to provide network security.

The subnet mask is rather like an IP address in that it contains four 8-bit words. It works by assigning a logic value of 1 to each bit in the IP address that is to be designated as network address, and a 0 to each bit that is to be designated as host address. So, taking the previous network address example 10.172.2.0, the binary equivalent value of this address is 00001010 10101100 00000010 00000000. To make this a class C IP address the subnet mask would have a value of 11111111 11111111 11111111 00000000. The dotted decimal equivalent to this would be 255.255.255.000. When overlaid on the IP address, we have:

class C IP address = 00001010 10101100 00000000 00000010 = 10.172. 2.000

class C subnet mask = 11111111 11111111 11111111 00000000 = 255.255.255.000

So we can see that wherever the subnet mask value is 255, the corresponding part of the IP address is designated as network address, and wherever the value is 000, the corresponding part of the IP address is designated as host address. Notice that because all eight bits of the last octet are designated as host, there will be $256 - 3 = 253$ possible host addresses available.

Following this to its conclusion, a class A IP address would have a subnet mask value of 255.000.000.000, and a class B subnet mask has a value of 255.255.000.000.

Now let's see how the subnet mask may be used to divide a network into four equally sized segments. Staying with the example 10.172.2.0, we know from Table 11.1 that we will have 2 097 152 possible network addresses with 253 hosts on each network. Suppose we now add the following subnet mask to this network address:

IP address = 00001010 10101100 00000010 00000000 = 10.172. 2.000

Subnet mask = 11111111 11111111 11111111 11000000 = 255.255.255.192

This has the effect of releasing two bits from the host address to the network address, increasing the number of networks to 8 388 608. However, there are now only six bits left for the host addresses on each network, meaning that we may now only have $64 - 3 = 61$ hosts on each network. The question must now be asked, why would we want to do this? When would an institution ever feel that they would require over eight million network addresses? However, the reason for subnetting is not to increase the number of available networks, but rather, to deliberately reduce the number of host addresses on each network. The reason why we would want to do this is because in practice the user may only have a few hosts connected to any one network, so the rest of the available addresses are unusable and are thus wasted. By breaking the original address down into four, with each subnet only having 61 host addresses, there is less chance of having unused addresses, and the other groups of 61 host addresses can be used on other networks that have been allotted the other subnets. Also, by reducing the number of hosts on each network, we further reduce the network traffic and therefore improve the efficiency.

Subnetting is carried out by the IT administrators and may be applied or modified at any time. The implications for CCTV are that, once cameras and other network devices have been set up in relation to any current subnets, changes to these subnets without modifying the network settings for each CCTV device will immediately cause them to drop off line. In practice, the CCTV engineer who has been called out to discover the reason for the sudden loss of CCTV images is rarely told about such changes and can therefore spend a lot of time looking for the cause of the trouble. A wise engineer would consult the IT department at an early stage in his investigation of the fault. Alternatively, he may check the current subnet mask against either historical records or the settings in the cameras/NVRs. Methods for performing this check will be discussed later in this chapter.

Assigning IP addresses

Every time a new host is connected onto a network it will need to have an IP address assigned to it. As we have seen, this address must be in the range for the class or subnet that is being used on that particular network, and it must not already be in use by another host or else there will be an *IP conflict* on the network. When this occurs, the new host will fail to communicate either with the other hosts, or the router/proxy server that is acting as the gateway to other networks. It is clear that to manually assign IP addresses would mean a lot of work for the IT administrators. Furthermore, on a large network, it would be very difficult for them to keep a track of the IP addresses, especially when devices such as laptops are constantly being connected, removed and moved around on the network.

Although it is possible to manually assign an IP address to a host, most networks have a *dynamic host configuration protocol* (DHCP) server which will perform this function automatically. DHCP functions as follows.

Remembering that a host connects to the network via its NIC, and that every NIC has a unique MAC address, the moment that a host is connected to a network, it begins to broadcast its MAC address. This broadcast is picked up by the DHCP server (assuming that one is available), which immediately assigns an IP address to that NIC/host. The DHCP server also issues a lease time on the IP address, for example, eight days (Figure 11.9a). The host will continue to use that IP address for as long as it remains connected to the network, but after 50% of the lease time has expired, the host will automatically request a renewal of the lease time, which will normally be granted. If the DHCP server is not available when the renewal is requested, the host will continue to use the IP address until 87.5% of the lease time is expired. It will then attempt to renew the lease time again, only now it will accept a lease from *any* DHCP server that responds.

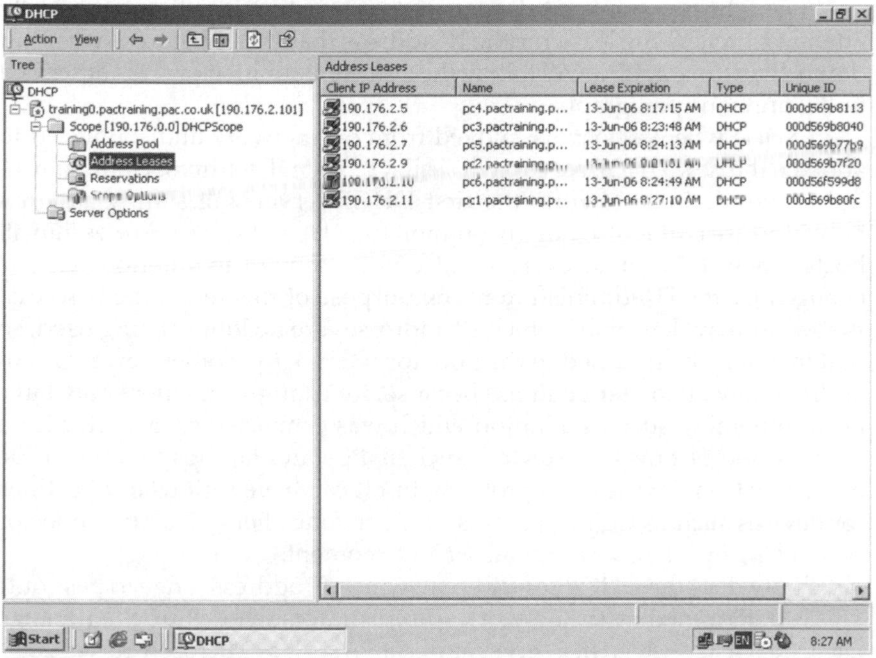

Figure 11.9a *IP addresses and lease expiry dates in DHCP. The right-hand column shows the MAC address for each host. This particular screen is taken from Microsoft Server 2000*

If the renewal is successful, the process simply repeats when 50% of the lease time has expired. If the host is unable to obtain a lease renewal, when the lease time is expired the host must discontinue using the IP address, forcing it to begin the process of obtaining a new IP address. If, for any reason, the host cannot contact a DHCP server to obtain a new address, after a short period of time it will assign itself one of the reserved autoconfiguration IP

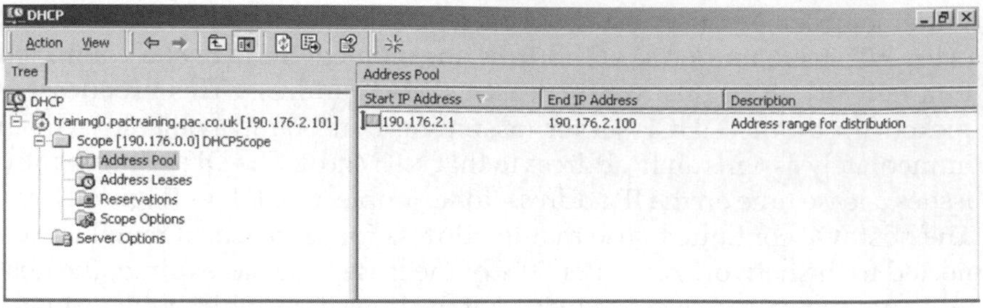

Figure 11.9b *Setting the IP address scope (range) in DHCP*

addresses in the range 169.254.0.1 to 169.254.255.254. The host will then attempt to contact a DHCP server every five minutes in an attempt to obtain an IP address that is valid for that network.

When a host is powered down, on re-connection it will automatically attempt to renew the lease for the IP address that it had before it was shut down. If unsuccessful, it will follow the autoconfiguration process described in the previous paragraph.

In a case where a host is removed from one network and connected to another, the new DHCP server will assign its own IP address. Upon returning to the original network, the first DHCP server will simply assign a new IP address. It is also not uncommon for a DHCP server to re-assign all hosts a new IP address every eight days, although this period may be changed by the IT administrator. The purpose of this re-shuffle is so that the server may determine which IP addresses are no longer being used, so that they may be returned to the pool for re-issue to another new host. An address may no longer be in use because, for example, the host may have been something such as a laptop which was connected by a visitor for a short period of time. For devices such as PCs and laptops this re-shuffle every fourteen days is not a problem, but it can have serious implications for devices such as CCTV cameras, IP door controllers, IP alarm collector points and such like, as we shall see in a moment.

So how does the DHCP server know which IP address range to pass out? This is determined by the network administrators who set up the IP range using the DHCP screen in a server application such as Microsoft Server 2003. A screen shot for the DHCP address pool is shown in Figure 11.9b. DHCP is also where the network administrators will set up subnets, should they choose to use them. The DHCP server can also manage gateway addresses, assign DNS servers (see later in this chapter), and perform many other tasks.

Manually assigned IP addresses

It was stated earlier that it is possible to *manually* assign an IP address to an NIC/host, but why would we want to go to this trouble? Well, one problem with using dynamic IP addresses with security devices is that these IP addresses may have to be manually entered into the software application that is used to administer the security devices; in the case of CCTV cameras, this is fine until DHCP changes the IP addresses of the cameras. When (or perhaps, if) this happens, unless the PC/NVR running the administration software is able to resolve a DNS name for each camera (see later in this chapter), it is possible that we may simply lose all communications with the IP cameras. Whether or not communications are actually lost will depend on the ability of the camera to provide a host name for DNS, the availability of a DNS server or, where NetBIOS is used, whether the network has been set up to resolve host names. Where there is a real possibility of communication failure caused by a change of IP address, the manufacturer will specify the use of static IP addresses for their equipment. Manually assigned IP addresses are referred to as *static* or, sometimes, *reserved* IP addresses.

To assign a static IP address to something like an IP camera we first of all need to know which address we are going to assign. There is no point in assigning an address that is already in use as this will simply result in an IP conflict on the network and a failure of the camera to communicate. Where the cameras are being connected to an existing LAN (that is, they do not have their own dedicated network), the only person who can provide the IP address for each camera is the network administrator. The IP address scope in DHCP may be set up such that a number of addresses are excluded from the normal DHCP range. These excluded addresses may be manually assigned to hosts requiring static IP addresses such as print servers, network printers and security devices such as IP cameras, door controllers, etc. Because these addresses are included in the DHCP scope, they will still be recognized for DNS purposes.

The next step is to assign the IP address to the camera, which is the tricky part because we cannot send the address via the network using TCP/IP since the camera does not yet have an IP address – a bit like the chicken and egg situation. Different manufacturers have different solutions to this problem. The simplest way to resolve this issue is to let the camera pick up an address from the DHCP server the moment that it is connected to the network, and then use proprietary software to probe the network for the cameras. The cameras will then reply using their dynamic IP addresses, at which point the commissioning engineer can use the software application to replace the dynamic addresses with static addresses.

The one drawback with this method is that it assumes that there is a DHCP server on the network, which is not always the case. Suppose, for example, that the IP cameras were to reside on their own private LAN; there would be no need for a DHCP server apart to assist in the initial set-up, which is a somewhat expensive luxury. And yet, without an IP address, how can the commissioning engineer communicate with the cameras to assign static IP addresses? The solution is to communicate using a protocol other than TCP/IP. Remember that every NIC or other network device has a unique MAC address, so it is possible to communicate using this in the first instance in order to assign the IP address, after which it can communicate using TCP/IP. In practice we are only doing manually what DHCP does automatically.

Address resolution protocol (ARP)

This is the method most commonly employed for communicating using the MAC address. It is the same type of broadcast that the NIC sends out when first connected to the network when it is looking for a DHCP server, except in this case the broadcast is being sent from the administration PC, prompted by the engineer. ARP performs a number of important functions on the network, all of which are beyond the scope of this text; however, for our purposes we can view the ARP broadcast as basically a message that travels over the network which says, 'Hey, MAC address 00:20:4A:32:F2:3C, your IP address is 192.176.2.13'. Being a broadcast, this request is heard by all hosts, but only the host with that unique MAC address will update its IP address. This is illustrated in Figure 11.10 where, in this instance, camera 1 would respond.

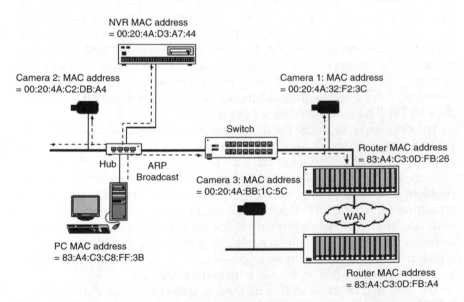

Figure 11.10 *Every network device that incorporates an NIC has a unique MAC (hardware) address. The ARP broadcast sent out by the PC would not pass beyond the router*

Once communications have been established, the engineer may assign the IP address to the device. Of course, he will need to know the MAC address in the first place in order to enter it into whatever software application he is using to perform the operation, which is why the MAC address is usually written on the item of equipment.

In practice the engineer will often need to do more than tell the camera its IP address. Bearing in mind that the engineer is performing the task of the DHCP server, he may also need to assign the gateway IP address (if there is a gateway between the camera and the NVR/admin PC) and possibly the subnet mask. Screen shots for typical utilities used to perform these functions are shown in Figure 11.11b and c.

One problem that may be encountered when using an application such as that in Figure 11.11b is that it may not be possible to connect to the remote device because of the presence of a router in the network path. Remember that the initial connection is being made using an ARP broadcast and, as we saw earlier in this chapter when discussing routers, a router will not pass a broadcast. This is illustrated in Figure 11.10, where an ARP request to connect to 00:20:4A:BB:1C:5C (camera 3) would be blocked by the router. The engineer, therefore, would be able to configure cameras 1 and 2, but would receive a 'Failed to connect to host' message every time he attempted to connect to camera 3. Note that in this example, camera 1 is connected via a switch, which means that it is on the same subnet as camera 2, so configuration is possible via a broadcast.

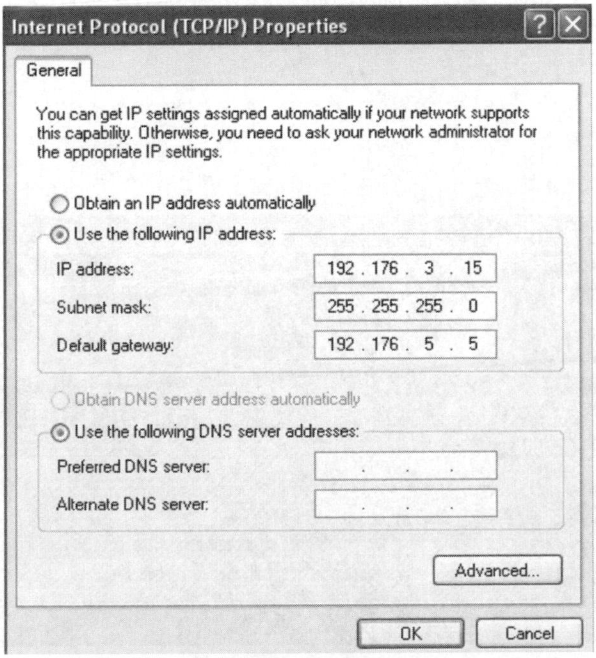

Figure 11.11a *TCP/IP set-up screen from Microsoft Windows XP, used to set up a static IP address on the PC itself*

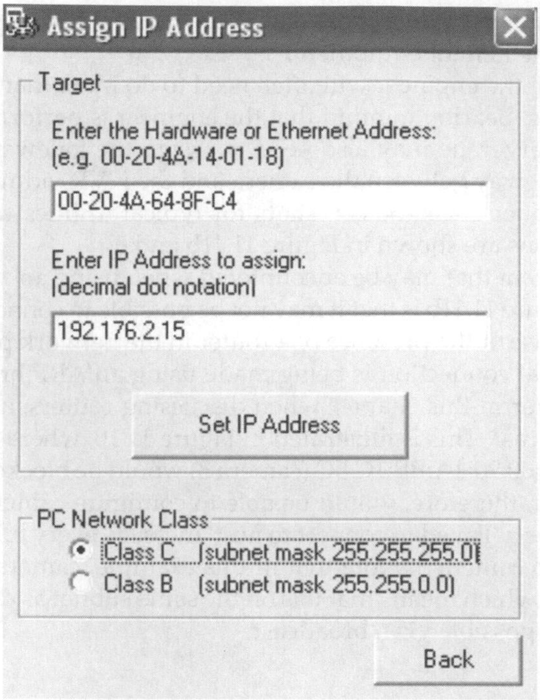

Figure 11.11b *A typical proprietary utility for assigning a static IP address to a device. Such utilities normally use ARP broadcast to communicate with the remote device*

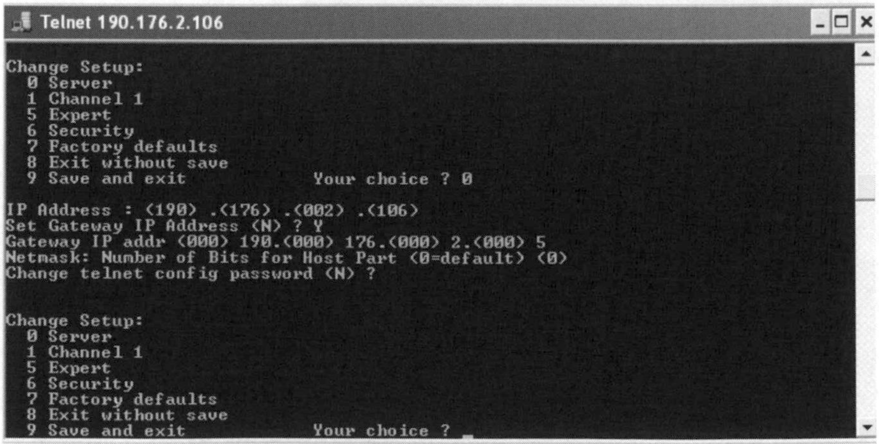

Figure 11.11c *Typical Telnet software utility used to set up a remote network device such as an IP camera, once an IP address has been assigned using ARP*

The problem is that in a large premises, the commissioning engineer often has no way of knowing the network topology, so one way around this problem is to preconfigure each security device local to the administration PC by connecting it via a hub or directly to the network port on the PC using a network crossover cable. Once configured and able to communicate using TCP/IP, the device may then be installed at its remote location. An alternative is to install the device set-up application software onto a laptop and configure the device locally, again by connecting via a hub or crossover cable.

Autoconfiguration

The more sophisticated security devices such as NVRs usually have an operating system such as Microsoft Windows, Windows CE or Linux. Where this is the case, the absence of a DHCP server is not such a problem when trying to assign a static IP address. Windows-based network devices are configured so that, after a few minutes of attempting to locate a DHCP server, if no server is found the device will autoconfigure and assign itself a reserved IP address in the range 169.254.0.1 to 169.254.255.254. As long as the commissioning engineer is prepared to wait for a few minutes following power-up and network connection for this process to take place, then it should be possible to probe for the device using the manufacturer's setup software, bearing in mind that routers should no longer present any problem because communication is now via TCP/IP. Once communication has been established, it should then be possible to assign the static IP address.

Domain name service (DNS)

One problem with using MAC addresses and IP addresses is that these long numerical strings are not very easy for humans to remember. For example, every website has an IP address, but imagine trying to remember the names of all of your favourite websites by using their IPv4 addresses, and those that use the more recent IPv6 addresses would prove impossible because they are 128 bits long!

To make networks user (human) friendly, another server is employed which looks at the IP addresses and resolves them into user friendly domain names. This server, known as the DNS server, may be separate from the DHCP server, although quite often on smaller networks one server may be configured to perform both functions. Domain names are hierarchical, the top names being the familiar .com, .uk, .ac, etc. The names that we would be concerned with when setting up network security devices are lower down the hierarchy.

Every host is given a name, usually by the network administrator, although in the case of devices such as IP cameras, the domain name normally incorporates some form of alphanumeric ID which the engineer may be able to modify during commissioning. When the DHCP server assigns the IP address to a new host, it also tells that host the IP address of the DNS server. Normally the host will then pass on its domain name to the DNS server, which will add this to its DNS record. Now, when a person working on another host wishes to communicate with that particular host (normally through some form of browser such as a web browser) that host will, in the first instance, only know the domain name that the person has typed in. Therefore it will put out a DNS request, and the DNS server will resolve (convert) the domain name into the IP address of the host to which they wish to connect. This means that users are able to work with user-friendly domain names rather than IP addresses.

Looking again at Figure 11.9a, the DNS names for the six hosts in that screen shot are listed in the second column.

There is another (older) name resolution service that is still used, called *Windows Internet naming service* (WINS). This service was designed to work on simpler networks (like the majority of domestic networks) where there is no DHCP or DNS server. These networks generally use a protocol called NetBIOS (also known as NetBEUI), which is simple to set up and administer; however, it is only suitable for small networks because it cannot be routed and, in terms of bandwidth, is very inefficient. Where CCTV cameras are running on their own private network it is possible that they are using NetBIOS, in which case there is every probability that WINS is employed for domain name resolution.

Note that WINS cannot be employed on any network that is intended to connect to the Internet because TCP/IP relies on DNS for address resolution. On the other hand, both WINS and DNS can co-exist on the one network.

Ports

For TCP/IP, a port is an address which defines the association between the data being transmitted and the applications on the source and recipient PCs for which the data is intended. For communication to take place, two ports are required, one at the source device and one at the recipient. The source port identifies the application that sent the data, whilst the recipient port identifies the application for which it is intended. Each port is assigned a number, some of which are well known because they are associated with common applications. For example, web servers use the protocol HTTP which has been assigned port 80, some email servers use a protocol known as POP3 which uses port number 110, and FTP servers communicate on port 21.

IP cameras must communicate via a port, the port number being dependent on how the images are being extracted. If a standard web browser is used, then communication will be via port 80, whereas an NVR running its own proprietary software would have a port assigned by the manufacturer.

A PC that is connected to the Internet may have any number of ports opened, depending on the applications that are installed. It is through these open ports that the all too familiar viruses, trojans, worms, adware, etc., come in. Those who perpetrate such annoying and often destructive software can set up their PC to 'sniff' hosts that are connected to the Internet to see if there are any open ports and then, upon finding a victim, the offending script is downloaded. *Firewalls* are used to combat this activity and they function by closing down all ports apart from those that are absolutely needed (port 80, for example). The problem for network CCTV is that, where there is a firewall between the cameras and the NVR or other image processing equipment, it will not be possible for the cameras to connect because their dedicated port will have been blocked. There is only one way to resolve this problem, and that is to have the network administrator open the required port on the firewall. However, in most cases they are reluctant to do this because it compromises their network security.

The problem of firewalls is unlikely to occur when all of the CCTV equipment is operating on one local network because there will be no need to connect via the Internet. However, more and more large organizations are wanting to network their CCTV systems nationally or even globally. Of course, it should always be possible to achieve this by using a web browser because port 80 is always open; however, the image quality is often not very good, and the frame refresh rate can be slow at times when the Internet is busy. In truth, the Internet was never intended to provide every large organization with a high-quality global CCTV system and such systems would operate far more effectively if private network connections were employed, but usually the cost of this proves to be prohibitive.

Other network protocols

TCP and IP are not the only protocols that may be used on a LAN or WAN, although in practice they are the most common. We have already looked at an example of another protocol in NetBIOS. Here we will briefly look at a number of other protocols that are used on modern networks.

User datagram protocol (UDP) is what is known as a connectionless protocol. When two hosts communicate using TCP, a session is opened between them, which means that a continuous connection is maintained until the end of the data transmission. By contrast, when UDP is used no such session is created. The datagrams are simply sent across the network to the recipient host where they may arrive out of order, and are not subject to any error-checking or correction.

On the face of it this all may sound very risky; however, it does depend on the importance of the data being sent and whether or not the application itself will initiate a re-send if no reply is received. For example, DNS uses UDP where, if the datagrams are lost, DNS will eventually try again.

The advantages of UDP are reduced data and increased speed, which are a direct result of omitting the session set-up and error detection.

Like TCP, UDP uses IP to send and receive the datagrams.

Some IP cameras employ UDP because of its clear reduction in network overhead, which is considered a worthwhile tradeoff for the occasional lost datagram which, in reality, represents only a very small portion of one TV frame. Where such cameras could begin to fall down is during periods of high network traffic, where collisions will be more prevalent and therefore more datagrams will be lost. But then it can be argued that cameras using TCP/IP will struggle to maintain image integrity during such periods, and their increased network overhead will actually exacerbate the network traffic problem.

File transfer protocol (FTP) is a part of the TCP protocol suite and is used as a means of transferring files between two hosts. Originally command line driven (i.e., using the DOS screen), user interface software applications now exist to simplify the use of FTP. To open an FTP session the user must know the IP address or domain name of the host with which they wish to connect and, depending on the host setup, they may have to enter a password before they are permitted to download or upload files.

FTP connections are made via port 21 although, once a connection has been established, a second TCP connection is set up on port 20 over which the actual data transfer takes place. The FTP connection is used for command and control data during the session.

Telnet is also a part of the TCP protocol suite and is a virtual terminal service which permits a user to connect two hosts and take full control over the remote host. All mouse and keyboard activity is passed directly to the remote host so that applications may be opened and run remotely. This may sound very grand, but one drawback is the latency in the system, especially when the network is busy, where keystrokes may take several tens of seconds to propagate through. Another frustration when using Telnet is its tendency to time out if the session is interrupted for too long when the network is busy. There are other proprietary software packages that perform the same function as Telnet, but with far less latency, and offering a far more reliable connection.

Telnet is, however, used by a number of security systems network device manufacturers to perform device set-up. Using a software application provided by the manufacturer, a Telnet session is established between the administration PC and the device to enable such things as subnet mask and gateway IP address to be programmed. An example of this is shown in Figure 11.11c.

Hypertext transfer protocol (HTTP) is well known to anyone who has used a web browser, even if they do not know exactly what it is. A typical web page comprises a mixture of text, graphics and links to other documents or websites. HTTP is a protocol that manages communications between the web browser and the web server and ensures that a requested folder, link, etc., is opened.

HTTP forms part of the *universal resource locator* (URL), for example, http://www.sitename.com. When considering CCTV network equipment, the URL may look something like http://devicename:urlpath. Sometimes the URL contains https rather than http. The 's' indicates that an encrypted communication channel must be set up between two hosts, which usually results in a prompt for a name and password. In the case of devices such as cameras and NVRs, the password is generated and established between the hosts automatically during initial network connection. This explains why devices that have been removed from one site and installed on a completely different site fail to communicate despite all attempts by the engineer during the commissioning stage. Most devices have some means of restoring factory default conditions so that a new password can be established between the new hosts.

Simple mail transfer protocol (SMTP) is responsible specifically for sending and receiving email between hosts. As it does not play any direct role in network CCTV, we will not discuss this any further.

IPv6

At the time of development of TCP/IP in the early part of the 1980s, no one could have predicted the unprecedented growth of the Internet and the number of network hosts that would be in existence today. Whereas the original concept was to have a large number of computers communicating on the network, today we are looking towards having a multitude of other devices such as PDAs, mobile phones, wireless network devices of all types, building management equipment, and domestic appliances such as refrigerators and heating/air-conditioning, to name just a few. This explosion in demand for network access leaves us with a very serious shortage of IPv4 addresses which, at best, can only provide $2^{32} = 4$ billion (approx.) IP addresses.

The issue is not just the shortage of IP addresses. As network sizes increase across the globe, the IPv4 system is becoming less efficient due to the large size of the routing tables required to maintain information about the network.

To satisfy the future demand for IP addresses, IPv6 has been developed. Unlike IPv4 addresses which comprise 4×8 bits $= 32$ bits per address, IPv6 addresses comprise 8×16 bits $= 128$ bits per address. This makes provision for $2^{128} =$ over 340 billion, billion, billion, billion different IP addresses. Working on the basis that there are approximately 6 billion people living on the planet, then each person could have around 56 billion, billion, billion IP addresses to use for their own personal items of network equipment, so we should have enough to be getting on with for a while!

An IPv6 address looks nothing like the familiar IPv4 addresses. Having eight sets of four 16-bit (hexadecimal) digits separated by colons, an IPv6 address looks like:

6AF4:0230:005A:0000:0000:0000:2D7A:437C

To simplify both the writing down and entering of these addresses, some rules have been applied that facilitate a possible shortening of the address. First of all, it is not necessary to show leading zeros in each group, so the address above could be written:

> 6AF4:230:5A:0:0:0:2D7A:437C

Secondly, a group or chain of groups that contain all zeros can be replaced by a double colon. So now the above address can be written:

> 6AF4:230:5A::2D7A:437C

There is one limitation to this second rule, which is that the address may only contain one double colon. Therefore, in an address that contains two or more separate groups of zero, only one group may be removed.

The migration from IPv4 to IPv6 is not simple and will take a number of years to implement. During the transition period, hardware must be capable of functioning on either network to facilitate the introduction of IPv6-ready equipment onto existing IPv4 networks.

Network diagnostics

When hosts fail to communicate on a network, there can be any number of causes, and it is up to the engineer to logically work through each possibility until the precise cause is located. However, logical fault diagnosis is only possible when the correct tests can be applied, but unfortunately the multimeter and oscilloscope are of little use in resolving communication problems on a network. There are network diagnostic tools available, but these are more likely to be used by dedicated network engineers rather than security systems engineers who, when it comes to network diagnostics, are mainly concerned with first line fault diagnosis. The sort of tools required for this type of fault diagnosis are found in a PC rather than a toolbox.

To begin with, don't forget to check the obvious. For example, if an IP camera is failing to connect to its host PC, NVR or MUX, check such things like: Has an IP address has been assigned to the camera? Is the IP address that is assigned to the camera being used by another device? Is the patch cable between the camera and RJ 45 socket correctly wired and not a crossover cable? Is the RJ 45 socket actually connected to the network?

As well as making a visual check, these points can be checked using the *PING* (packet internet groper) utility. PING is normally executed from the command line (i.e., in the DOS screen) and it provides a simple means of testing if it is possible to reach a particular host on the network and obtain a reply from that host. When PING is executed, the PC simply sends out a message on the network which says, 'host so-and-so, are you there?'. If the message reaches the device, it will attempt to send back a reply which is then displayed on the PC. If a reply is received, it confirms that all hardware and connections between the PC and the host in question are good, that the host is able to communicate across the network, and that the IP address is valid on that network.

To execute the PING command you must first open the DOS screen, which is usually done from the Start button on the bottom toolbar (assuming the operating system is Microsoft Windows™ 95 or later). From the Start menu, select the Run option and, for Windows 2000 or Windows XP type 'cmd' (for Windows 95/98 type 'command') – the DOS screen should open with the flashing cursor at the CMD prompt. At this prompt, type 'ping' followed by either the IP address or the hostname for the device you are looking for – in this case the camera in question. A typical ping command will look like:

ping 192.176.5.10

A few seconds after pressing the Enter key, a response will be obtained. If a device is present, the response on the screen will look something like:

pinging 192.176.5.10 with 32 bytes of data:

Reply from 192.176.5.10:   bytes = 32   time = 10 ms   TTL = 128
Reply from 192.176.5.10:   bytes = 32   time = 10 ms   TTL = 128
Reply from 192.176.5.10:   bytes = 32   time = 10 ms   TTL = 128
Reply from 192.176.5.10:   bytes = 32   time = 10 ms   TTL = 128

If no device is present, the response 'Request timed out' will appear. Bearing in mind that, among other things, you are trying to ascertain whether or not there are two devices sharing the same IP address, disconnect the CCTV camera in question and re-execute the Ping command. If a response is now obtained you will know that you have been given an IP address that is already in use and the system administrator will need to provide you with an alternative. Another consideration when no reply is received is that the network administrator has suppressed ping commands on his network. Either ask him, or try pinging a known working host.

Another useful tool for finding out information about a PC or PC-based DVR that uses a Microsoft Windows operating system is the *ipconfig* utility. There are a number of variations, called switches, to the basic command; however, simply typing 'ipconfig' in the command line will reveal the IP address of the PC, its subnet mask and the default gateway IP address (if there is a default gateway). A typical response is illustrated in Figure 11.12a.

The '/all' switch provides much more information about the PC on which the command is executed. As shown in Figure 11.12b, in addition to the information derived from 'ipconfig', we can now see the physical (MAC) address, DHCP configuration, the IP addresses of the DHCP and DNS servers and the expiry date of the DHCP assigned IP address. This command is executed by typing *ipconfig/all* in the command line.

Two other switches that may be used with 'ipconfig' are '/release' and '/renew'. These can be very helpful if you have just changed the IP

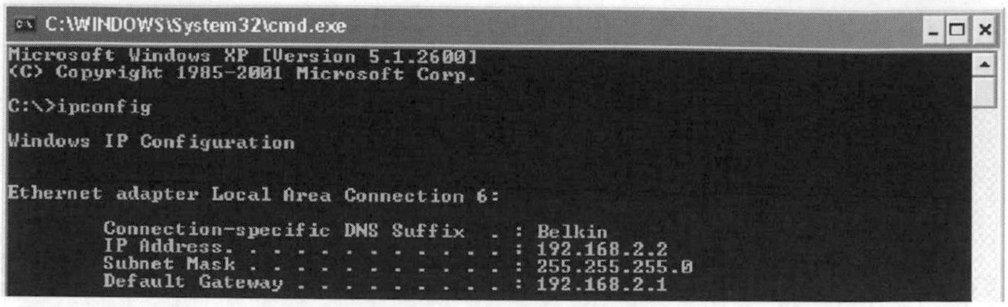

Figure 11.12a *Typical response to* ipconfig *on a PC*

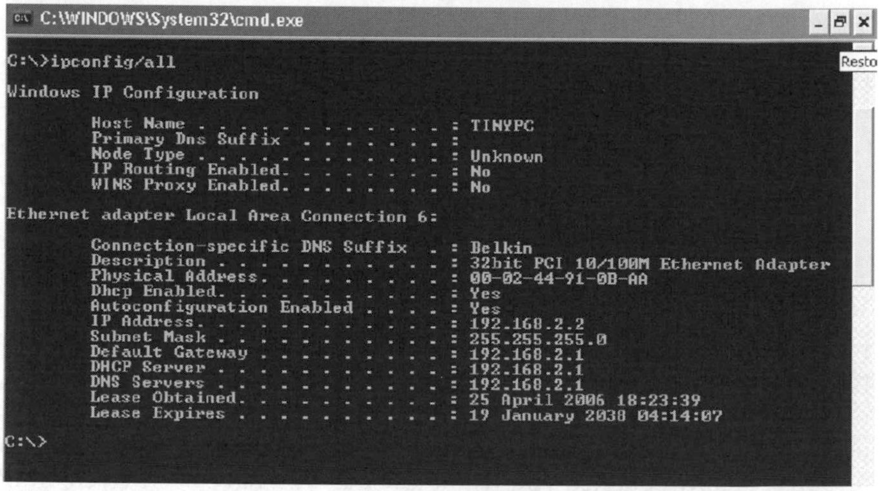

Figure 11.12b *Typical response to* ipconfig/all

address configuration of a PC from static to dynamic (see Figure 11.11a). Having assigned the dynamic address, the PC may not see the new address until it has been re-booted, which may take some time. Typing *ipconfig/release* in the command line forces the PC to release any existing IP address. If you then follow this by typing *ipconfig/renew*, the PC will look for the new (dynamic) IP address.

The *traceroute* utility may be used to discover the path from a PC to another host on the network, either local or across the Internet. Typing *tracert* at the command line produces a list of every router (including the IP address of each router) through which TCP/IP packets pass in order to reach a particular host. It also shows the time (in milliseconds) that it takes for the packets to pass through each router. An example is shown in Figure 11.13, where the path to the web server for www.pactechinfo.co.uk is shown. The trace reveals that the packets pass through twelve routers after hop number one which, as we can see from the information in Figure 11.12b, is not a router IP address but the address of the local DNS server.

```
Command Prompt                                               _ □ x
Microsoft Windows XP [Version 5.1.2600]
(C) Copyright 1985-2001 Microsoft Corp.

C:\>tracert www.pactechinfo.co.uk

Tracing route to pactechinfo.co.uk [83.170.69.95]
over a maximum of 30 hops:

  1     1 ms     1 ms     1 ms   192.168.2.1
  2     9 ms     8 ms     9 ms   10.10.32.1
  3    28 ms     8 ms    10 ms   oldh-t2cam1-a-ge-wan41-107.inet.ntl.com [80.5.16
4.29]
  4    11 ms    11 ms    37 ms   manc-t3core-1a-ge-011-0.inet.ntl.com [213.104.24
2.41]
  5    10 ms    16 ms     9 ms   man-bb-a-so-400-0.inet.ntl.com [213.105.175.1]
  6    10 ms    11 ms    11 ms   lee-bb-b-so-700-0.inet.ntl.com [62.253.185.194]

  7    24 ms    11 ms    10 ms   lee-bb-a-ae0-0.inet.ntl.com [62.253.107.105]
  8    46 ms    14 ms    13 ms   nth-bb-b-so-000-0.inet.ntl.com [62.253.185.101]

  9    17 ms    16 ms    48 ms   tele-ic-1-so-700-0.inet.ntl.com [62.253.184.2]
 10    16 ms    18 ms    16 ms   linx.uk2net.com [195.66.224.19]
 11    17 ms    18 ms    16 ms   gw12k-hex-g0-0-303.uk2net.com [83.170.69.234]
 12    16 ms    26 ms    19 ms   ultra25.uk2.net [83.170.69.95]

Trace complete.

C:\>_
```

Figure 11.13 *Typical response to* tracert *command. In this case the path from a PC to a website named www.pactechinfo.co.uk has been traced*

Earlier In this chapter we saw how Telnet may be used to connect two hosts in order to control one host from the other. However, you may also use this utility to test the connectivity of any port on the network, which can prove useful for checking if a host is capable of responding via a particular port. By typing 'telnet' in the START – RUN menu in Windows XP, a Telnet session may be opened. Typing 'help' in the command line reveals a list of all the functions within this utility, and how to execute them. To execute a port test on a host, type 'open hostname/ipaddress' [port number]. For example, typing 'open 192.176.2.2 [23]' will create a session on the host having that IP address, on port 23. Clearly, if the host name or IP address of a CCTV camera plus the port number used by that camera is known, then it will be possible to check to see if the camera is communicating on that port. It should be pointed out that, if the connection is successful, there will be no notification on the screen; however; if the connection is not successful, then a 'Connection failed' message will be displayed.

A failure to communicate with the IP cameras could, of course, be due to something as simple as a defective or incorrectly set network interface card on the administration PC. Where the PC is connected to a main network, it is easy to verify if this is the case by simply trying to connect to another host, such as a local file server or a well known and, therefore, reliable Internet site. However, where the CCTV system is operating on its own dedicated network, with no connection to other LANs/WANs, another means must be found to test NIC functionality and network set-up. One such test is the software *loopback* test, which is performed by entering, at the command line, the reserved IP address 127.0.0.1. This test does not actually produce any network traffic; however, the response (shown in Figure 11 14) does confirm that the NIC is functioning.

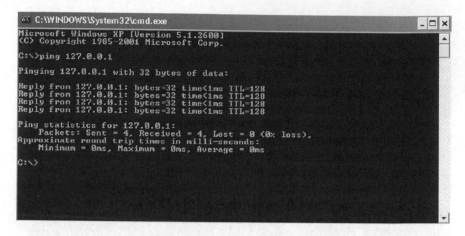

Figure 11.14 *Pinging 127.0.0.1 produces a similar response to any other ping; however, the responses are coming from the host's own network interface card*

CCTV over a network

In Chapter 2 we considered network cabling, examining the bandwidths and bit rates that are available from the different cable categories and the factors that may reduce those bandwidths. In Chapter 5 we looked at video compression techniques, and saw how the file size of each picture frame may be reduced to improve the efficiency of transmission over networks, as well as reduce the required storage capacity of digital recording equipment. We also saw how excessive compression may result in poor image quality due to inabilities of the decompression codec to recover all of the lost information. In Chapter 8 we looked at digital recording and the different options available for storage, retrieval, downloading and archiving of video information. Having now looked at the basics of networking and seen how devices communicate over a network, we are able to bring all of these factors together and consider the practicalities of using a network to transmit digital video signals.

The implications of utilizing networks for CCTV (and indeed security alarm and access control systems) are enormous. Where structured cabling is already installed in a premises, or where it is possible to utilize the existing network for CCTV, installation is greatly simplified because of the reduced cable installation. Furthermore, by utilizing the network, we introduce the possibility of remote administration of the system from any point on the globe. However we cannot, and must not, ignore the practical considerations.

First of all there is the issue of bandwidth. Even a compressed video signal requires a relatively large amount of network bandwidth, and the more cameras that we place on the network, the greater will be the bandwidth required. However, it is not always simply a case of knowing that there is sufficient bandwidth on a network to handle the video traffic. Where the cameras are operating on an existing LAN that already has a large number of network hosts, we have to consider the amount of *available* bandwidth. Where a network is already operating almost to capacity, the addition of just one camera may well take it over the top, resulting in intermittent loss of connection of some devices, slow network functionality, picture freezes on CCTV cameras, etc. In cases such as these, it may not even be the camera that goes off line, but other devices such as PCs, printers, scanners and, more important, servers dropping. In such cases it is often the CCTV engineer who is blamed for the problem because, when all is said and done, everything worked until the cameras were put onto the network!

As a general rule, when considering using IP cameras on an existing network, the specifier or engineer must first ascertain the limitations of the existing network and ask questions like: How much available bandwidth is there? What is the available bandwidth at peak times such as early morning and early evening when everyone is either logging onto the network, or is logging off? (Remember, if there is an IP access control system on the network, these times will also represent the peak traffic periods for the access control because everyone is swiping in or out of the building.) How many cameras will be required? What image quality does the user require? What frame update rate does the user require from each camera? What is the bandwidth requirement of each camera for this image quality and frame rate? And finally, how well has the network performed to date? Before connecting your CCTV system to an existing network, it is prudent to find out if there has been a history of recurrent communications problems, otherwise you may find yourself attempting to resolve problems that have nothing at all to do with the CCTV installation.

For larger IP CCTV systems where it is felt that an existing network would be unable to support the system, it is very common to have a dedicated network installed solely for the use of the CCTV system (and possibly other security systems such as intruder and access control). In these circumstances it is usually necessary to install new cables (unless there is sufficient spare capacity in the network cable bundles) but at least the CCTV system installer retains complete control of the CCTV IP network. He does not have to worry about the IT department changing subnet masks, adding firewalls between networks, altering the network security such that he is no longer able to make changes to the administration PC, or adding another bandwidth-hungry application such as voiceover IP.

And, of course, having a private network for the CCTV system does not mean that the images cannot be viewed from other parts of the network, or even the world. The CCTV network may still be connected to another LAN via a bridge, switch or router and, using suitable software and a password, users are still able not only to view CCTV images but interact with the system, moving cameras, stream video from NVRs, etc.

Network CCTV example

A typical network CCTV system is illustrated in Figure 11.15. In this example the CCTV system enjoys the luxury of having its own private CAT 6 LAN and, with only seven cameras on the network, provided that the network is installed to CAT 6 standards, the system will operate very efficiently. The NVR will constantly record the images streaming from the seven cameras whilst at the same time forward those images to the administration PC, which is also serving as the main viewing monitor. The only

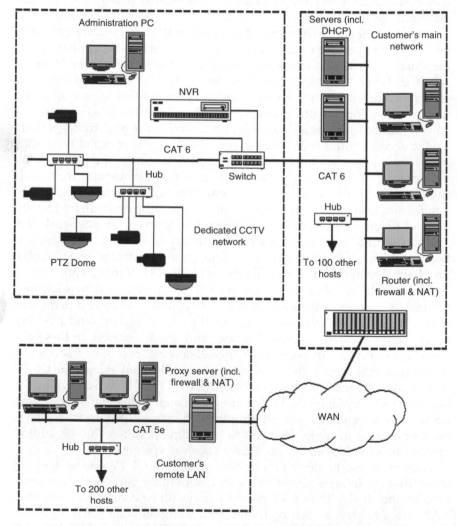

Figure 11.15 *Example of a CCTV network, connected to an existing network via a network switch*

other network traffic on the LAN will be the telemetry between the administration PC and the three PTZ dome cameras.

A network video recorder functions like any other network device in that it will have a MAC address, it usually has a static IP address assigned to it, it connects to the network via a NIC, and it uses either TCP or UDP to perform data transport. For the majority of the time an NVR is simply recording data (video images) from the IP cameras on the network, although it may also be providing a stream of multiplexed images to one or more viewing monitors, depending on the operational requirement for the CCTV system.

The network traffic in our example may be reduced if cameras having in-built motion detection are used. In this case, where there is no movement within the field of view for a predetermined time, the camera will cease to send its images across the Web. However, this does have one disadvantage over using motion detection in the NVR in that, when an alarm event occurs, the NVR may be unable to retain the video information prior to the alarm event because it may not have been transmitted by the camera.

One of the prime advantages of network CCTV is that the images may be viewed from anywhere in the world, provided that a suitable network link is available. There are a number of methods by which the images may be accessed, but not all of these are practical in all situations. Looking again at the example in Figure 11.15, the user would be able to view any of the camera images from any of the PCs on the local network using a standard web browser, provided that the IP address of each camera was known. Alternatively, with the appropriate manufacturer's software installed on a PC, the user would not only be able to view images but may actually control the PTZ domes. Another alternative would be to access the NVR and stream the most recently recorded images for any camera, thus in effect viewing almost live images without having to know the IP addresses of every camera.

Possible drawbacks of streaming images across LANs (i.e., via the network switch in the example in Figure 11.15) are, firstly, that the images may appear at the viewing monitor in a more highly compressed state than if they were viewed directly at the NVR. Secondly, if there is a lot of network traffic on the user's main network, the frame update rate may be much lower because of the time that it takes to receive all of the data packets required to construct each frame.

In Figure 11.15, the main network includes a DHCP server. This can simplify the initial setting up of the IP cameras by assigning a dynamic IP address the moment they are connected to the CCTV network. Once a dynamic IP address has been assigned, it is a simple task to manually assign a static IP address, because the camera is able to communicate using TCP/IP. In the absence of a DHCP server the camera will be unable to obtain an IP address, and the only way to assign one manually would be to use the ARP command, either directly or via a software application like that illustrated in Figure 11.11b.

Note that if a router had been used in place of the switch, the cameras may not be able to obtain an IP address from the DHCP server because, unless the router has been set up to do so, DHCP will not function across a router. Remember how DHCP operates. The moment that an IP camera is connected to the CCTV network, it will broadcast its MAC address in the hope that a DHCP server picks up the broadcast and, in turn, assigns a dynamic IP address. However, a router will block the broadcast, so the DHCP server will never receive it. It is possible to configure a router to pass the MAC broadcast in order for DHCP to operate, but if the network administrators have not done this, the IP cameras will have to have their static IP addresses assigned using the ARP command, even though there is a DHCP server on a near-by network.

Another way of overcoming the problem of DHCP being blocked by a router is to install a *relay agent* on the CCTV network. This device uses TCP/IP to communicate with the DHCP server and requests IP addresses and leases on behalf of the hosts on its network. So, the DHCP server is still issuing the IP addresses and leases, but it is doing so across the router by sending them not directly to the host, but to the relay agent using TCP/IP.

The example in Figure 11.15 includes a connection across the WAN to another LAN owned and administered by the same user. NAT devices are used to connect the two LANs and, as would be expected, these devices include firewalls. It will also be noted that the remote LAN is CAT 5e, which will be much slower than the CAT 6 local network, not only because of the lower specification, but also because of the number of hosts residing on the CAT 5e LAN.

Trying to view CCTV images on the remote LAN could prove to be problematic, although in practice it is usually possible when everything is set up correctly. First of all there is the issue of connecting to the CCTV network where, as when viewing images from the local LAN, the simplest method is to use a web browser to connect to the NVR. However, there will be a lot more latency in the system as a result of the action of the router and the proxy server, plus the narrower bandwidth and lower speed of the CAT 5e LAN.

Connecting to the NVR from the remote LAN using manufacturer's proprietary software will almost certainly be out of the question because the port will be blocked by both firewalls. Loss of this option will mean that the user will be unable to operate the PTZ domes from the remote LAN. Only by having the IT administrator open the ports used by the NVR on both firewalls will this method of communication be possible and, for good reason, the IT administrator is usually reluctant to do this as it will make the network more vulnerable to attack.

Integrating analogue cameras

The example in Figure 11.15 makes the assumption that every camera is a network camera which, in the case of a new installation, may be the case.

However, for the foreseeable future there will be a requirement to maintain existing analogue cameras and integrate them into network CCTV systems.

One possibility is to employ an NVR that accommodates analogue inputs as well as network cameras, which is all right as long as the analogue inputs terminate in the same room as the NVR is to be located, and as long as none of the camera locations requires telemetry.

An alternative solution is to use a device that digitizes and compresses the analogue video signal prior to transmitting it over a LAN or WAN. The principle is illustrated in Figure 11.16, where a four-way video server is depicted. Servers such as this simplify the task of integrating analogue and network systems by enabling the analogue cameras to be connected to the network at a location much closer to the camera. This may well mean a considerable reduction in the length of the co-axial cable, which in turn should provide improved image quality.

Video servers like that shown in Figure 11.16 are available in different forms, ranging from a single analogue input device that may actually be located inside the camera housing, to large rack mount arrays. The rack mount array would more than likely be located in the control room, which removes the advantage of co-axial cable length reduction. Nevertheless, the existing analogue cameras are still being made available on the network, and may be accessed from a much wider area. A typical rack employs separate four-way servers, making the system completely expandable. Each individual server has its own RJ 45 network connector, which means that a network switch would also be required in order to connect each of these to the LAN.

Figure 11.16 illustrates another advantage of employing video servers, which is that PTZ telemetry signals may be passed via the Ethernet, through the server and on to any PTZ or dome cameras via their co-axial connections. In order to achieve this, a server with telemetry capability must be used, and it must have appropriate drivers installed in order for it to produce the correct protocol. In most cases drivers for the majority of the popular PT/dome cameras are available, and the required driver may be downloaded from the administration PC to the server.

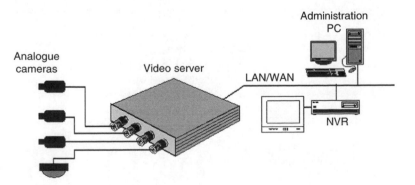

Figure 11.16 *Four-way video server converting four separate analogue video signals into a TCP/IP network signal*

Summary

In this chapter we have examined the basic operating principles of Ethernet networks. By no means can we hope to cover the detail of this subject in such a small chapter, bearing in mind that most networking textbooks are at least 5 cm thick! However, what we have done is to focus on the issues that are particularly relevant to security systems engineers who are having to install and set up network security devices such as IP cameras, alarm collector points, access control door controllers, and the like.

Network CCTV is by no means going to completely replace analogue systems in the short term, and there is a firm role for both technologies for the foreseeable future. Nevertheless, it is perhaps as short-sighted for security systems engineers to ignore network technologies as it would have been for shorthand typists to ignore the advent of the word processor. As network performance and technologies continue to improve, security systems in general will continue to take more advantage of networks. Network CCTV images will continue to improve, frame refresh rates will become faster, even at peak network traffic periods, and more reliable methods will be derived for extracting and archiving video evidence over a network. For all of these reasons it is imperative that security systems engineers have a grasp of network technologies, just as they have had to acquire a grasp of PC hardware and Microsoft Windows operating systems.

12 Ancillary equipment

Up to this point we have examined each of the primary components in a CCTV system, looking at their operating principles, their function, variations in technology, setting up and adjustment methods, and have identified typical fault symptoms and causes. In this chapter we shall deal with those components in a CCTV system which, although at times may appear mundane, nevertheless play an essential role.

Camera mountings

CCTV cameras are both expensive and delicate and in many cases need to be protected from the elements, vandals, thieves, or a combination of these threats. A range of protective mountings are available and the choice for any given application may be determined by such factors as the degree of protection required, size of camera, aesthetic requirements, mounting location and whether it is to be a covert or overt installation.

The simplest form of camera mount is a wall bracket and, although not offering protection against any threat, in indoor locations where the threat is minimal, brackets offer an effective and inexpensive means of fixing a camera. All cameras have a standard mounting, which somewhat simplifies the choice of bracket, and the only points left to consider are the length and weight of the camera. For obvious reasons weight is an important factor, but it is not only the load-bearing ability of the bracket that must be considered – the surface to which the bracket is to be fixed must also be taken into consideration. For example, where an internal camera is to be fixed to a thin plasterboard surface, it may be advisable to employ a toggle or gravity bolt fixing; however, not all internal brackets have mounting holes large enough to accommodate such fixing devices. The length of the camera must be taken into consideration when choosing a bracket to ensure that there is enough clearance behind not only to allow free movement when setting the camera angle, but also to provide sufficient room to allow for the cabling.

Some brackets, both internal and external, incorporate a cable management system where the video and power cables pass through the bracket. This is an important feature where cables are prone to vandalism or malicious attack, and this feature also improves the aesthetics of the installation.

When selecting a suitable housing for a camera the engineer should consider such points as internal or external use, the size of camera, the method of fitting and removing the cover(s), wash/wipe and heater requirements, overt or covert design, the degree of risk of attack (i.e., should an anti-vandal type be used?) and whether there any ergonomic requirements. The traditional housing design incorporates an elongated shell with an option for P/T mounting or a wall-mounting bracket, perhaps incorporating integral cable management. The metal casing should be protected from corrosion by galvanization, anodizing or coating with a weatherproof paint. There will be a footplate inside the housing onto which the camera will be fixed and, in the case of housings intended for mains voltage operated cameras, power terminations at the rear of the assembly. When used in cooler climates, external units should have a thermostatically controlled heater to prevent misting of the front glass and the lens. These heaters usually take the form of a 15–20 W wirewound resistor connected in series with a thermistor to the mains supply voltage. For anti-vandal versions the front glass will be shatterproof.

The shell will have some means of opening to allow camera installation and future maintenance and for this there are a number of design trends. For larger housings, the top cover is generally held in place by clips and lifts off for access, allowing ample room to work on large camera/lens assemblies. One problem that may be encountered is what to do with the housing top once it has been removed. If working from a hydraulic platform there should be room to place it on the floor, but if the engineer is only required to perform a simple operation which may be accessed via a ladder (which, in the UK, is often not permitted under the regulations for working at heights), then grappling with a large housing cover may become something of an issue. In some cases the manufacturer provides for this eventuality by fixing a steel wire between the cover and the housing bottom, allowing the top to hang from the assembly when removed.

Smaller housing designs have a variety of access methods but some of these may prove problematic for certain installations. For example, in some cases the camera is fixed to a footplate which is then inserted into the rear of the housing. In theory the engineer should be able to swing the housing around to give room for this operation, but this is not always possible. Furthermore, once installed, it is very difficult for the engineer to then perform adjustments to the lens and back focus. Other designs have the top cover slide on and off which, again, is not always a simple operation in some situations. For most situations the best form of housing cover is where the top is either hinged, or is lifted off and allowed to hang from a secure cable. For higher security situations where the camera might be accessed by unauthorized persons, securing the cover using anti-vandal screws is a viable option.

Another design feature is the provision of a sun visor. Direct sunlight on the lens may cause the iris to close down, resulting in a very dark image, or it can cause multiple reflections within the lens that produce undesirable light streaks or halo effects on the picture. Similarly, when sunlight falls directly onto the front glass, unwanted haloes or bright patches may appear on the picture. To eliminate these problems, external housings should incorporate some form of sun visor. This is generally formed into the top cover and provides shielding to both the top and sides of the front glass.

Another factor that may determine the choice of external housing is whether or not a wiper is required, as not all housings are designed to accommodate this. The wiper and associated motor are often specified by the manufacturer as an optional item, allowing the installer to only use these where it is felt necessary and therefore reduce the installation cost. When considering these costs it is important to remember that it is not only the wiper hardware that must be included but also the cost of providing telemetry to control the motor. In the case of a fully functional camera this is not a problem because telemetry will already be present; however, for fixed cameras where there is no need for telemetry the added cost of a wiper plus telemetry will raise the cost of that camera installation considerably, begging the question: does that particular camera really need a wiper facility?

A wash facility may be installed at camera locations where there is excessive dust or dirt, but careful consideration must be given to this. It will be necessary to accommodate a reservoir somewhere close to the camera in a place where it will not be tampered with. In the example shown in Figure 12.1 the reservoir is housed in the base of the camera column, but this will not be possible on a wall-mounted camera installation. Wash reservoirs designed for location at the camera head are available, but in many cases it may prove very difficult to replenish them. In the case of wall-mounted installations, some installers have attempted to locate the reservoir on the roof of the building; however, because it is above the outlet nozzle the water is prone to siphon away. In the majority of cases a wash facility is not essential because a well-maintained wiper will be able to cope with light deposits of dust and dirt, assisted often by the action of the rain.

Dome housings have rapidly overtaken the more traditional fixed and pan/tilt (P/T) housings like that in Figure 12.1, partly because they are much more pleasing aesthetically, but also because the fixed versions offer easy 360° camera positioning, and the powered versions offer much more rapid 360° panning movement than their traditional counterparts (this was discussed in Chapter 10). Another advantage of the dome housing when used externally is the fact that it negates all of the problems of wind drag which, as we shall see in a moment, affect housings mounted onto P/T units. Perhaps one drawback with the external dome housing when compared with a traditional design is that, at the time of writing, no-one has come up with an effective design for a wash/wipe mechanism other than a mop on a long pole! In particularly dirty environments this should be given serious consideration because, where the camera is not easily accessible, the lack of a wipe facility can mean that the pictures very quickly become obscured by splashes of dirt on the dome.

Dome housings are particularly useful where it is desirable not to reveal the direction in which the camera is pointing. By coating the dome

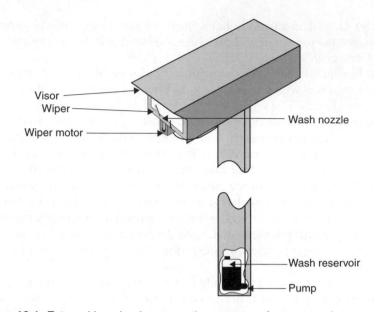

Figure 12.1 *External housing incorporating common features such as wash and wipe and sun visor*

with chrome or aluminium, or by using a smoked cover, the camera can be concealed. The drawback of using such dome covers is that the light input is attenuated – by as much as two f-stops in the case of a coated cover and as much as one f-stop in the case of smoked covers. A further drawback is that the curvature of the dome cover will inevitably introduce a degree of optical distortion, and in some cases this can be quite noticeable. When choosing a dome the issues of light attenuation and optical distortion should be considered.

Although some manufacturers of camera housings simply state the intended applications for each of their units, the majority use Standard BS EN 60529: 1992, which is the correct method of specifying any form of enclosure within the electrical industry. The degree of protection afforded by an enclosure is indicated by a rating known as an *index of protection* (IP). The IP codes are summarized in Table 12.1. In the code, the first digit indicates the level of protection against ingression by solid objects, whilst the second digit indicates the level of protection against liquids. Where an 'X' appears in place of the first or second digit, this indicates that there is no guarantee of any protection in the corresponding area. An additional letter may be placed at the end of the IP number in cases where it is necessary to indicate a level of protection of persons that is higher than that afforded by the first digit.

As can be seen, the codes are intended to include a wide range of protection ratings ranging from minute particle ingression to whole body ingression, and from full immersion in water to no moisture protection

Table 12.1 *Outline of IP (index of protection) codes*

IP number	First digit (touch & solid objects)	Second digit (liquids)
0	No protection	No protection
1	Protection against contact by large part of body or objects >50 mm diameter	Vertical dripping water shall not impair equipment operation
2	Protection against contact by Standard Finger and objects >12 mm diameter	Dripping water shall not impair equipment operation when housing is tilted to 15° from vertical
3	Protection against solid objects >2.5 mm thick	Spray or rain falling 60° from vertical shall not impair equipment operation
4	Protection against solid objects >1 mm thick	Spray or rain falling from any direction shall not impair equipment operation
5	Protection against dust ingress in quantities capable of interfering with equipment operation	Low pressure water jets from any direction shall not impair equipment operation
6	Total protection against dust ingress	High pressure water jets from any direction shall not impair equipment operation
7	–	Immersion of housing shall not permit water ingression (within specified time & pressure)
8	–	Indefinite immersion of housing shall not permit water ingression

Additional letter: A = Protection against contact by a large part of the body
B = Protection against contact by a Standard Finger
C = Protection against contact by a tool
D = Protection against contact by a wire

whatsoever. From the point of view of CCTV this may appear somewhat extreme; however, it must be borne in mind that these ratings apply to any containment intended for electrical installations, including high voltage.

Example: A camera housing has a rating of IP50. In this case the housing would offer a high level of protection against dirt and dust, but absolutely none against moisture. Such a housing would be suited to indoor use only.

Example: A camera housing has a rating of IP66. In this case the housing would offer a high level of protection against dirt, dust and moisture. Such a housing would be highly suited to outdoor use.

Example: An electrical equipment housing has a rating of IP4X. In this case the housing would offer a reasonable level of protection against small objects but give no guarantee regarding moisture ingression. Such a housing would be suited to indoor use only, and even then only in a clean and secure environment.

Example: An electrical equipment housing has a rating of IPXXA. In this case the housing would offer protection against accidental human contact but give no guarantee regarding moisture ingression, and does not protect against dust, dirt and deliberate ingress by persons. Such a housing would be suited to indoor use only, and even then only in a clean and secure environment.

Towers and columns

Mounting a CCTV camera on top of a tower is common practice, but there are a number of things to take into consideration such as positioning, possible planning permissions, installation of supply and signal cables to the site, the physical installation of the tower, cable management, protection against attack, installation of site drivers and other ancillary devices and access for servicing. Let's look at each of these in turn.

Positioning of a CCTV camera tower can be very important and the specifier who looks at the problem purely from the point of view of obtaining the best field of view may find him/herself coming unstuck. For example, companies that have installed cameras directly in front of a private dwelling may quickly find themselves having to disconnect them when the owner of that dwelling takes out a court order on the grounds of it infringing his/her human rights. It may also be that the location of the tower is determined by the location of existing underground cable ducts, the cost of excavating for new cables being prohibitive. In areas where building development is still taking place, the specifier must try to take into account the erection of structures that may impair the camera once it has been installed; it would not be the first time that a lamp post or hoarding has been erected directly in front of a CCTV camera!

Town councils are very particular about what can and cannot be erected in their centres, and special planning rules may apply; for example, the style of the tower may have to blend with that of the street lighting and, in the UK, most local councils will only permit the use of dome housings in town or city centres because of their aesthetic qualities.

The laying of underground cables is always expensive, so if a cost-effective means of laying the power and signal cables out to a camera site can be found, it is always worth considering even though it may not be the best location from a field of view perspective. The client has to decide if a marginal improvement in camera performance is worth a considerable increase in installation cost.

Tower structures require a solid anchor and should be installed by skilled civil engineers. Manufacturers will provide guideline instructions on such things as the minimum depth and area of the foundation, but knowing the correct mix and the minimum curing time of the concrete is something that only qualified persons will know, not to mention that digging holes in built-up areas often unearths such things as high-voltage cables, sewers, and gas and water mains.

CCTV camera towers are generally of two designs: the solid column structure and the lattice tower (Figure 12.2). The lattice tower is by far the strongest structure because its low surface area offers less wind resistance

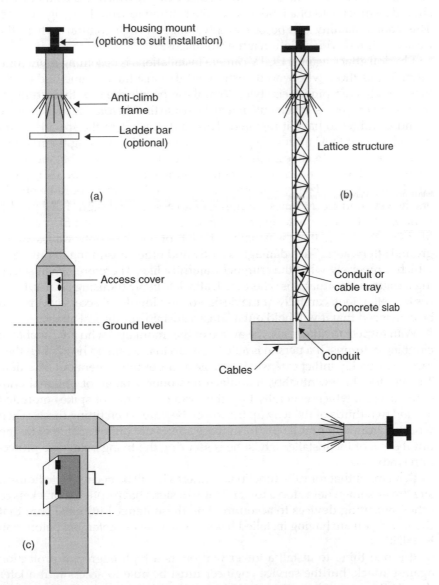

Figure 12.2 *Typical camera towers and columns. (a) Fixed column – may be cylindrical or hexagonal. (b) Lattice tower structure – may also have a wind down mechanism. (c) Wind down column*

and the triangular lattice produces a very rigid construction. However, an open lattice means that the cables must pass largely unprotected up through the centre of the tower, a feature that is not very desirable for most installations. Some protection can be afforded by using steel conduit, but this will not prevail against a determined attack. Solid columns are, by comparison, far more suited to the task because they offer considerable protection of the vertical cables. With regard to their strength and rigidity, a hexagonal solid column will be more stable than a circular design, which means that there should be much less of a tendency for the column to wobble in high winds. This added stability can be of real advantage in high columns where the camera is fitted with a high-magnification lens.

Physical attack against CCTV camera installations is becoming more commonplace as those who would wish to indulge in a life of crime find CCTV increasingly cramping their style. What these people are rapidly learning is that once a camera is 'taken out', it may be some time before the owners have found a budget to have it repaired. This said, it is up to the specifiers and installers to do their utmost to protect cameras, and mounting them atop a high tower is one effective method of achieving this, provided that the access cover at the base of the tower is secure! Manufacturers have for a long time employed anti-vandal bolts or other methods of deterring would-be attackers; however, in recent times a number of access cover designs have proven to be woefully inadequate against a large crowbar, and once a fire has been lit inside the tower there is usually little left of the fibre-optic cables above ground afterwards. Such damage is costly and time-consuming to put right, which is precisely what the criminal fraternity like. Therefore, when selecting a column for situations where such attack is likely (housing estates, town centres, etc.), look carefully at the design of any low-level access covers and be convinced that it will hold out against a sustained physical attack.

With respect to attack, also be aware of the 'monkeys' who are capable of climbing columns. If a person is able to get up to the camera head, then they can very quickly inflict extensive damage and put the camera out of action. To cater for this eventuality, manufacturers offer a range of optional anti-climb devices which generally take the form of a series of spikes mounted around two thirds of the way up the tower. Beware when fitting these; when working from a hydraulic platform these spikes are quite capable of taking out the eye of the installer whilst he is securing the fixing bolts. Eye protection is advised.

Remember that for fully functional cameras it will be necessary to house a site driver somewhere atop a tower. Many designs have optional brackets or other mounting devices to accommodate these items, but make sure that the tower you are having installed has such facilities – preferably before you install it!

It is one thing to install a tower that offers a high degree of protection against attack, but the service engineer must be able to access it at a later date. In the case of lower towers having only a fixed camera, ladder access may be adequate, but this does mean that a ladder bar should be fitted to the tower before it is erected. These bars usually drop down over the tower and are then secured. The bar offers a solid point against which a ladder can rest and be secured, and a well-designed bar will have flanges, one each end, to prevent the ladder from sliding off. This said, when working in the UK, engineers must always be mindful of the regulations that apply when working at heights; and in many cases working from a ladder, even if secured at top and bottom, would not be acceptable.

For higher towers with more complex installations on top, access will either have to be via a hydraulic platform, or a wind-down tower may be installed. There are a number of designs of winding mechanism available, depending on the weight of the tower. When installing such towers, take into consideration the direction in which it must wind down; for example, having it fall across a motorway carriageway might not be the best idea! Also, be aware that you may require special training in the use of the winding mechanism before being permitted to operate it. In recent years there have been a number of serious accidents (some fatal) where engineers have failed to operate the winding mechanism correctly and either the winding handle has spun out of control, hitting the operator, or the tower has come down too fast and struck the operator or other persons close by.

Pan/tilt units

Earlier in this chapter it was pointed out that a P/T dome housing has many advantages over the more traditional pan/tilt unit and camera housing assembly. Nevertheless, the older style camera housing still has some advantages over dome technology; for example, dome housings do not easily accommodate large lenses, cleaning can be problematic, a dome assembly does not easily accommodate lighting units, and motorized dome assemblies, at the time of writing, still tend to be more expensive than equivalent P/T and housing assemblies. There are still many occasions when these advantages tip the choice of assembly in favour of a camera housing which, when the camera needs to be moved, necessitates the use of a P/T unit.

Where it has been decided to use a traditional P/T unit and camera housing, before selecting the unit there are a number of points to consider such as loading, wind drag, supply voltage, maximum drive speed, and whether the assembly is to carry lighting.

Loading refers to the physical weight that the pan/tilt unit is to carry and will include the combined weight of the housing, camera, lens, lighting units and anything else that is on the assembly. A pan/tilt unit with an excessive load will initially fail to meet operational expectations, its movement being sluggish and possibly having a tendency to overrun when the operator expects it to stop. However, in the long term the excess load will cause wear of the gears and possibly motors, resulting in premature failure of the unit. A further problem that may be encountered with excess loading is that the unit may have insufficient power to lift the assembly up when it has been tilted downwards beyond a certain point.

Each pan/tilt unit has a maximum load rating which is typically between 12 kg and 50 kg, although with the reduction in size and weight of cameras and lenses during recent years the requirement for heavier duty units is becoming less. When selecting a unit the specifier should first of all calculate the total weight of the proposed assembly and then, as a rule of thumb, add another 2 kg allowance for cable drag and wind resistance. In environments where snow or ice are likely to build up on the housing, a further 3 kg should be added to the total weight calculation. Having arrived at a final figure for the maximum possible load, a unit with a rating that exceeds this figure by at least 15% should be selected. Another consideration with regard to loading is future developments. If it is anticipated that further equipment may be added to the assembly at a later date (for example, a pair of lamp units), it is advisable to include the weight of this equipment in the initial calculation to save having to replace the pan/tilt unit with a higher specification when the system is upgraded. Where infrared lighting is a requirement, one possible method of reducing the load on the pan/tilt unit is to use LED lamp units rather than tungsten halogen types (see Chapter 3).

In open areas the problem of wind drag can be quite serious. An under-rated pan/tilt unit will not be able to function well in strong winds, and it is not uncommon to have badly specified assemblies that simply spin around like a weather cock on a church spire in strong winds because the forces acting on the housing overcome the torque in the gear train. In some cases this action can break the teeth off the drive gears, although better-quality pan/tilt units may incorporate some from of clutch assembly which will allow them to slip without damage. Having said this, a correctly rated and installed pan/tilt unit should function perfectly well in strong winds, although the units may require more frequent servicing in the form of lubrication and greasing than units that operate in less hostile conditions.

Some pan/tilt units have the housing mounted on top whereas others have it mounted to one side (Figure 12.3). In general, top-mounted units tend to have a lower load capacity than an equivalently rated side mount unit because of the difference in the centre of gravity of the load with respect to the drive mechanism. This point is illustrated in Figure 12.4 where it can be seen that, because of the leverage effect of the housing on the drive shaft, the amount of energy required to lift the assembly back from a tilted position is much greater for a top mount pan/tilt than it is for a side mount. This does not mean to say that side mount units are necessarily better, because a top mount design should always be able to perform to its design specifications; it is just a fact of physics that a side mount unit will be able to handle its load with a lesser amount of energy than its top mount equivalent.

Top mount units come into their own when the assembly is required to have one or two lamp units fitted. Although it is possible to hang a lamp unit from the underside of a side mount assembly, it is far easier to do this with top mount units – especially when two lamps are required because in this

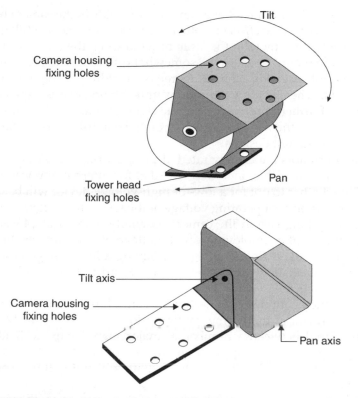

Tilt

Camera housing
fixing holes

Pan

Tower head
fixing holes

Tilt axis

Camera housing
fixing holes

Pan axis

Figure 12.3 *Top and side mount pan and tilt units*

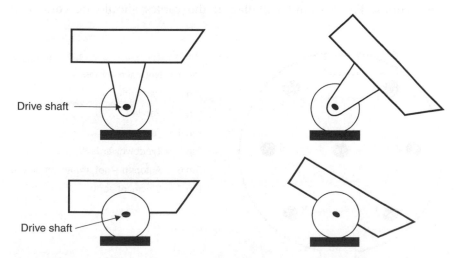

Drive shaft

Drive shaft

Figure 12.4 *A side mount unit does not have the problems of leverage associated
with top mount units*

case they may be mounted below and to either side of the camera housing, producing a balancing effect on both the assembly and the pan/tilt gear box.

Returning once more to the issue of wind drag, the effects of wind on top and side mount units can be somewhat different. The pan/tilt unit in a side mount assembly offers a degree of shielding to the housing from side winds, whereas for a top mount unit the housing is exposed in all directions. Furthermore, because of the leverage effect of the housing on a top mount unit, the wind forces acting on the pan/tilt gears are far greater than for side mount designs.

Pan/tilt motors are usually rated at 24 Va.c. or 230/110 Va.c., although d.c. units are available, including 12 Vd.c. for internal applications. The decision whether to opt for a low- or high-voltage device will be dependent largely upon the operating voltage of the rest of the system. For example, if everything else in the camera assembly is rated at 24 Va.c., there seems little point in employing a high-voltage pan/tilt unit as this would negate one of the main advantages of having a low-voltage system, i.e., simplified installation requirements with regard to Electrical Installation Codes of Practice.

The methods of power connection to pan/tilt units can differ between manufacturers, but the wiring configuration given in Figure 12.5 is one that is shared between a number of major manufacturers. Before purchasing a unit for an application, make certain that the connection arrangement is suited to the site driver with which it will be expected to function.

Cable management for fully functional camera assemblies is very important because a poorly installed connection between the moving housing and the fixed site driver can result in restricted movement, additional loading, water ingress and failure of the assembly due to broken connections. For a sound installation the cables should be contained

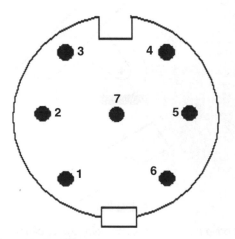

Pin 1 = Neutral/common

Pin 2 = Pan right

Pin 3 = Pan left

Pin 4 = Tilt upwards

Pin 5 = Tilt downwards

Pin 6 = Autopan – not always employed

Pin 7 = Electrical earth

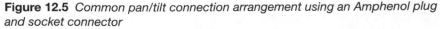

Figure 12.5 *Common pan/tilt connection arrangement using an Amphenol plug and socket connector*

within a flexible conduit and fixed at both ends using appropriate moisture-resistant glands. All cable entries should be at the bottom of both the housing and the side driver containment to minimize the chance of moisture ingression. The cable should be long enough to permit free movement of the assembly through all of its required field of view and should neither restrict nor be restricted by the movement of the housing. Having a long drop of cable between the site driver and the housing is the key to unrestricted movement and limited loading effects.

A particular point to look out for is where the housing assembly is fixed at the end of a boom arm which is itself secured to a solid structure – usually a building. In this case the cable should be secured by some form of tie wrap along the boom before entering the housing, allowing enough slack at the housing end to permit free movement without the flexible conduit rubbing on the boom. Such a rubbing action will result in wear of the conduit and subsequently the cables inside.

The drive speed for a pan/tilt unit is quoted in degrees per second, and this can vary considerably between different units. Panning requires less energy than tilting, so for any given unit the panning speed is usually much higher than the tilting speed. When considering quoted speeds, bear in mind that these are generally quoted assuming a near to maximum load condition; however, this speed performance may alter if the load figure deviates either side of the maximum stated load condition.

To prevent a pan/tilt unit from overrunning in either the horizontal or vertical directions, limit switches may be employed. For a d.c. motor circuit this principle was discussed in Chapter 4 with regard to the zoom lens motor and is the same for the pan and tilt motors in a P/T unit. When the pan/tilt unit reaches the limit position, the switch opens and power to the motor is cut. As can be seen from Figure 12.6, diodes across the switches enable the motor to power-up when the polarity of the supply is reversed. For a.c. motors the principle is somewhat different because it is not possible to reverse the motor by simply reversing the voltage polarity. In this case the limit switches are connected to the site driver, which will cut power to the motor. Motor reversal is then usually performed by having two sets of windings in the motor. Limit switch adjustment generally involves some method of releasing microswitches and sliding them to the desired positions before securing them once again. A minimum of four switches is required to set limits for left, right, up and down movement. Limit switch adjustment is important if the housing is to be prevented, for example, from hitting a wall or from tilting downwards to a point from which the motor has insufficient torque to lift it back up.

Pan/tilt units require routine maintenance if reliable performance is to be assured. This largely takes the form of lubrication or greasing of the gears but in some cases may involve the cleaning away of old grease that has become thick due to ageing effects or contamination from dirt or metal particles from worn gears. During maintenance, gears should be inspected for signs of wear and replacements fitted as necessary. Where the gears have worn away, or where the teeth have been stripped by the action of

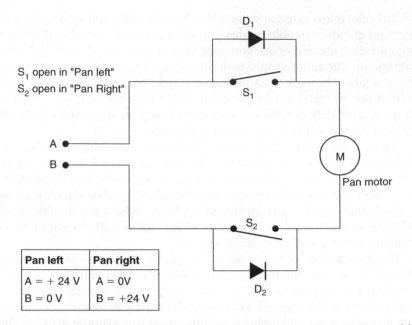

S₁ open in "Pan left"
S₂ open in "Pan Right"

Pan left	Pan right
A = + 24 V	A = 0V
B = 0 V	B = +24 V

Figure 12.6 *Limit switch circuit for a pan motor. The same arrangement would be used for the tilt motor*

Figure 12.7 *A P/T unit where the tilt mechanism gears have failed. This is a fairly common problem with these units, especially if the unit is carrying an excessive load*

high winds or excessive loading, the P/T unit will either spin around freely (if the pan gears have failed) or fall down to a vertical position (where the tilt gears have failed). A typical example of such a failure is shown in Figure 12.7. It should be pointed out that any servicing that requires the unit to be dismantled would normally be undertaken in a workshop rather than at the camera head.

Motors are prone to overheating when the movement of the unit becomes restricted, which in turn may be caused by faulty gears or overloading of the unit. A motor that has overheated will often have short circuit turns in its windings which will result in a reduction of torque, causing the unit to become sluggish. In extreme cases a motor may cease to function altogether, resulting in the loss of either the pan or tilt function. Where a motor is found to be defective, it is worth investigating the possible causes for its failure rather than simply installing a replacement.

Other servicing points would include a visual inspection of all cables, conduits, mounting brackets, etc. for signs of wear/corrosion. Also check that limit switches are secure and are still functioning.

Monitor brackets

In many respects the same rules that apply to the selection of camera brackets also apply to monitor bracket selection; however, it must be kept in mind that a CRT monitor is generally much heavier than an internal camera installation. There are a number of swivel mount bracket designs available specifically for monitor use and it is advisable to employ these rather than adapt something else, if only to avoid litigation in the event of any mishap. But of course, it is still up to the installer to ensure that the bracket chosen is rated to carry a load in excess of the expected monitor weight.

The installer must also verify that the surface to which the bracket is to be fixed is capable of retaining the proposed fixings. Where there is any doubt, additional means of support must be applied – monitors are generally sited where people circulate, and one simply cannot take the risk of a unit falling onto a person.

Power supplies

It is usually a simple thing to connect all of the equipment in the control room to the 230/120 V mains supply. However, this is not always as easy for cameras and other remote equipment because it may involve a lot of expensive civil work to install the cables. Alternatives to mains-operated cameras are 12 Vd.c. and 24 Va.c. versions. These use appropriate low-voltage power supplies which may be installed either in the control room or at suitable locations around the site. Extra low voltage (ELV) cables are installed alongside the co-axial signal cables to carry the power.

A variety of 12 V and 24 V power supply units are available offering different current ratings, typically between 1 A and 4 A. As with any power supply, the required rating is determined by the load, and care must be taken not to over-run a power supply as it will very quickly fail. The issues surrounding power supply rating for cameras were discussed in Chapter 6.

An a.c. power supply unit is primarily a mains to 24 V step-down transformer with a rating of at least 1 A. For higher current ratings a larger transformer may be used, but this can lead to problems with voltage regulation. When a transformer is operating with little or no load, the secondary voltage often rises above that stated, the output voltage falling progressively as the load current is increased. It is asking a lot of a transformer to provide a constant 24 V across a load current range between 0 A and 4 A, and because of this some larger a.c. power supplies incorporate a number of separate 1 A rated transformers. Alternatively a transformer with a number of secondary outputs may be used.

A d.c. power supply is more complex because it requires rectifier, filtering and regulation circuits (see Figure 12.8). The voltage regulator has the task of maintaining a constant 12 Vd.c. output irrespective of load current. However, it will have a maximum current rating, usually between 1 A and 4 A. Overload protection is essential if damage is to be prevented in cases where the rating is exceeded or an accidental short circuit occurs across the line. A fuse is the simplest form of protection; however, many regulator ICs incorporate an overload protection circuit which switches the output voltage to 0 V until the overload condition is removed.

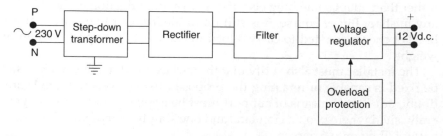

Figure 12.8 *Block diagram of a d.c. power supply*

Power supplies are generally constructed within metal housings to prevent external RFI getting onto the supply lines, and also to contain any EMI coming from the power supply. This screening only functions effectively if the earth connection to the casing is properly made.

The main advantage of using ELV is the fact that cable installation is very much simplified and avoids the more involved inspection and testing required for 230/120 stallations, which include visual inspection of all mechanical protection and housings, insulation resistance tests, earth loop impedance tests, and polarity checks. In essence, mains installations must be carried out by a 'competent person', which does not simply mean someone who thinks they know what they are doing, but someone who has proven their competence – in other words, a qualified electrician. By contrast, in the UK, 12/24 cables only have to comply with IEE regulations regarding segregation and mechanical protection, and they must be tested for earth leakage and insulation resistance integrity prior to connection into the system.

Voltage drop

The main problem associated with low-voltage supplies is that of voltage drop, especially in longer cable runs. For any given conductor, if the cross-sectional area is doubled, the area through which electrons pass is doubled, so the resistance is halved. Conversely, doubling the length of the conductor will double the resistance. This relationship is expressed in the formula:

$$R \propto \frac{L}{A}$$

where the symbol $\propto$ means 'is proportional to'. This relationship is illustrated in Figure 12.9.

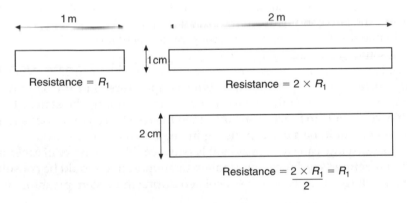

Figure 12.9 *Effect of conductor length and area on resistance*

All cable contains resistance, and although this is very small in a short length, because the resistance increases with conductor length, for longer cable runs the resistance can amount to a few ohms. It should also be remembered that a power supply requires two cables for positive and negative or, in the case of an a.c. supply, send and return. Therefore the resistance is doubled. The equivalent circuit for a power supply and load showing the cable resistance is given in Figure 12.10.

For d.c. and 50/60 Hz a.c. supplies the cable can be considered to have a purely resistive effect, and thus voltage drop can be considered from the point of view of Ohm's Law. For the circuit in Figure 12.10, the voltage drop would be found from $V = I \times (R_i + R_{ii})$, where the current can be considered to be the rated current for the equipment constituting the load.

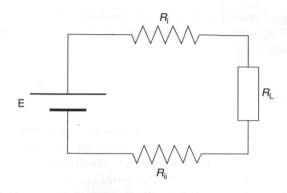

Figure 12.10 *Cable resistance in a circuit, represented by* R_i *and* R_{ii}

For example, if the load is a colour camera for which the manufacturer has specified a current of 300 mA, then the voltage drop would be taken to be $V = 300 \times 10^{-3} \times (R_i + R_{ii})$.

Cable resistance is usually quoted in Ω/m or Ω/km; however, where this is the case, the resistance figure usually refers to the total series resistance of the two conductors in the circuit (i.e., $R_i + R_{ii}$), so there is no need to double this figure when performing cable calculations.

In some cases a manufacturer might quote actual voltage drop figures, in which case you might see something like: 80 mV/A/m at 22°C. Note that the resistance of copper increases with temperature, and this is why the temperature at which the measurement was taken must be stated. If this cable were made to function at 30°C, then the resistance would rise slightly and there would be a corresponding increase in the voltage drop.

The resistance can vary considerably between different types of cable, and a manufacturer's technical information or support line should be consulted where voltage drop needs to be calculated during the system planning stage.

Example

The camera in the system illustrated in Figure 12.11 is quoted as having a current rating of 250 mA and a minimum operating voltage of 9 Vd.c. The resistance of the cable is quoted as being 0.03 Ω/m at 20°C. Calculate the voltage drop in the cable, and determine whether or not the camera would be able to operate satisfactorily.

Solution

Total cable length = 450 m
Total cable resistance at 20°C = $450 \times 0.03 = 13.5\,\Omega$
Voltage drop = $I \times R_{cable} = 250 \times 10^{-3} \times 13.5 = 3.375\,V$
Voltage at camera = $V_{psu} - V_{cable} = 12 - 3.375 = 8.625\,V$

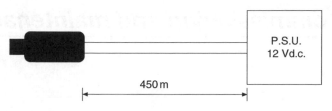

450 m

Figure 12.11

Thus it can be seen that the camera would not function correctly, if at all. One possible method of overcoming this problem would be to use a higher-gauge cable, remembering that increasing the cross-sectional area reduces the resistance. Where a multicore cable is being used with spare cores available, doubling up the cores can cure the problem. In the case of the above example, doubling the cores would reduce the cable resistance to $6.75\,\Omega$, and therefore the voltage drop would be $250 \times 10^{-3} \times 6.75 = 1.7\,V$. This gives $12 - 1.7 = 10.3\,V$ at the camera, which is within its operating voltage range.

It is sometimes stated that the voltage drop for a.c. is 'less than for d.c.', and that this is why a.c. power supplies are better. While it is true that a.c. supplies are better because the voltage drop appears to be less, in fact the $50/60\,Hz$ alternating current is subject to just the same resistance, and hence power loss, as a d.c. current along the same cable. The reasons why voltage drop figures appear to be less for a.c. powered cameras are: (1) the current consumption of the camera is less because of the higher operating voltage (24 V); (2) the 24 V is actually the RMS figure, and the true peak value is $24 \times 1.414 = 33.9\,V$. Summarizing this point: the voltage drop for a.c. power supplies is less of a problem because the reduction in current causes a reduction in voltage drop; and a voltage drop of, say, 3 V equates to an RMS loss of only $3 \times 0.707 = 2.12\,V$.

13 Commissioning and maintenance

Commissioning

System commissioning is vitally important. This is the point in the installation where every part of the system is tested against the original specifications to ensure that they are met in every way. It is where the system may be proven to be meeting any operational requirement (OR) that is in place (in terms of such things as live and recorded image resolution, camera fields of view, zoom capability, and ability to perform under differing lighting conditions). It is where the quality of the installation workmanship is proven. It is where the safety of the system is proven (in particular, does it conform to the IEE 16th Edition of the Wiring Regulations – BS 7671?). It is where test results are recorded for future reference and, when commissioning is properly carried out, it is where the installation company can hand the system over to the client with confidence.

To begin with, a visual inspection of all parts of the system should be carried out. In particular the connections to fully functional cameras should be checked for freedom of movement and, in the case of external cameras, integrity of weatherproof seals. All cables should be clearly tagged and identified.

Programming and setting up of equipment should be checked to ensure that the system will meet the OR, or where this does not exist, that it will meet the customer's requirements and needs. The inspection should include the programming of multiplexers, matrix switchers, operator levels of authority (restricted access), PTZ limit switches, dome presets, DVR/VCR time-lapse settings, alarm input response(s), and VMD zones and sensitivity.

The correct operation of all equipment must be confirmed, not forgetting peripheral items such as lights, video replay facilities, alarm detectors, video printers, and bulk tape erasers.

It should go without saying that the results of all of these tests and inspections should be documented and filed for future reference. The installing company should also notify relevant authorities such as the local council (where planning permission was originally required for external equipment installation), police, Inspectorate body, etc.

Measuring resolution

The subject of resolution has been discussed on a number of occasions in this book, but how can this actually be measured in a way that is meaningful to all concerned? Or to put this another way, how can the commissioning engineer actually prove that the system is performing to the design specifications, or operational requirement? Resolution is affected by many factors, and one cannot make assumptions about the quality of the picture, even when the system has been built using high-specification equipment throughout. Changing lighting and weather conditions can have a dramatic effect on system performance, as can other changes in the environment in which the cameras are operating. For example, in a town centre system when the Christmas illuminations are turned on, the dramatic change in the colour temperature can cause iris settings to alter, which in turn may cause changes in the depth of field and focusing, not to mention the fact that some cameras may be blinded almost completely.

Various methods have been devised for testing the picture performance of a system, perhaps the simplest of these being the use of an approved test card. The card is positioned in front of the camera, which is adjusted so that the image fills the entire monitor screen. In this condition each set of markings on the card corresponds to a particular TVL picture resolution, which is normally specified in the instructions that come with the test card. Note that the monitor should be adjusted for best picture before any observations are made.

To test the system performance under site conditions the test card should be held or fixed in front of each camera in turn, and the camera adjusted so that the image of the card fills the monitor screen. The maximum TVL resolution is determined from the smallest set of markings that can be resolved on the displayed card. It is not uncommon for the measured resolution figure to be lower than that quoted for the camera or monitor. This may be due to such factors as lighting conditions, length of cable runs, or the quality of other components in the system. For example, a high-quality camera fitted with a poor-quality lens will have its performance impaired.

An important point to note from the test procedure described above is that it applies to a situation where a stationary image is filling the height of the screen. But how does this relate to an image of a person filling only, say, 25% of the screen height, and moving at speed? This is a much more subjective measurement and has been the cause of much debate.

In the UK, the Home Office Scientific Development Branch (HOSDB, formerly PSDB) has carried out research into the issue of image size and picture resolution, and from this research a set of guidelines has been devised. These have been adopted by inspectorate bodies such as NACOSS and the SSAIB.

The HOSDB have formulated the following classifications of CCTV system use *monitoring, detection, recognition* and *identification.*

Monitoring is defined as an image that allows the observer to see the location, speed and direction of a person in the field of view. This will usually be a wide-angle view. For monitoring purposes the image of a person will be not less than *5% of the screen height.*

Detection should allow the observer to locate a person with a high degree of certainty having been prompted to do so by a guard, police, alarm system or other means. For detection purposes the image of a person will be not less than *10% of the screen height*.

For *recognition*, the picture quality must be adequate to permit an observer to say with a high degree of certainty that the person on the monitor is the same one they have seen before. In this case the image of a person will be not less than *50% of the screen height*.

Identification is the highest resolution image of a person and must contain sufficient detail to enable the operator to see the person clearly enough to be able to describe them, or to identify them again. Such an image is only possible from a close-up or zoom shot, the disadvantage of this being that you cannot see or record any activity other than that by the person being monitored. For identification purposes the image of a person will be not less than *120% of the screen height*.

These image sizes are illustrated in Figure 13.1. Note that for each classification it has been assumed that all parts of the system are adjusted and are functioning correctly, and that the height of the person is 1.6 m. It might also be necessary to test the performance under different lighting conditions.

From these specified image sizes it is possible to draw up system specifications which include a TVL figure for a given image size. For example,

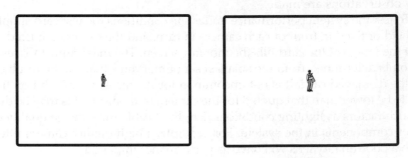

Monitoring
Image = 5% of the screen height

Detection
Image = 10% of the screen height

Recognition
Image = 50% of the screen height

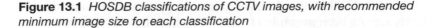

Identification
Image = 120% of the screen height

Figure 13.1 *HOSDB classifications of CCTV images, with recommended minimum image size for each classification*

the specification for a certain camera location could be that it will be used for recognition purposes, and that the image must have a resolution of at least 250 TVL. In other words, the image of a person filling 50% of the screen height must have a resolution of 250 TVL. For an engineer to test this during final commissioning using a test card, the camera will be adjusted so that the card fills 50% of the screen height. Under these viewing conditions, the 250 TVL markings on the card should be discernible.

An alternative to using a test card is to employ a test target known as a *Rotakin*, illustrated in Figure 13.2. This target stands 1.6 m tall when fixed to its frame and, in addition to the human head outlines, has a range of markings relating to TVL. The TVL figures for each marking are given in the table

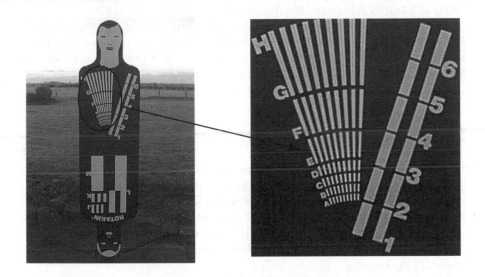

Marking	TVL figure for 100% R
A	500
B	450
C	400
D	350
E	300
F	250
G	200
H	150
J	100
K	80
L	40
M	20

Figure 13.2 *Rotakin standard target. The higher resolution bars are highlighted in the close-up picture*

accompanying Figure 13.2. Note that these figures apply when the image height of the Rotakin fills the monitor screen, a condition defined as 100% R.

Up to this point we are still looking at a stationary, clear target, and we have yet to take into account a camouflaged moving target.

To meet the problem of camouflage the Rotakin comes with a combat-style jacket which can be fitted when required. This is particularly useful when testing the system performance for picking out people other than black and white striped ones! In other words, in this mode the Rotakin is no longer being used to determine TVL resolution, but rather the ability of the system to display a useful image under difficult conditions. With the jacket fitted it is also possible to estimate the response time of the operator.

To simulate a moving target the stand on which the Rotakin is mounted has a 12 V motor which rotates the target at a rate of 25 rpm. This provides a representation of a person 'moving quickly but stealthily', to quote the Rotakin manual. The original function of the rotating target was to measure the amount of lag in the image, which is an effect that is produced by tube cameras; however, because tube cameras are no longer used, the Rotakin is generally only ever used as a stationary target.

System handover

Handover is one of the most important stages in a CCTV installation, and yet it is too frequently glossed over as something that is only of limited importance. It has been proven that where a well-presented handover has taken place there is much less chance of multiple-site re-visits during the first few months of the system going live. Such visits are far more likely to occur when the owner/operator has little or no idea of how to operate and maintain the system.

Handover is the point where the engineer becomes a training officer, and as soon as he/she realizes this, the whole issue of instructing the customer takes on a new importance. Have you ever attended a manufacturer's training course where the person delivering the training appears covered in dirt, wearing overalls that have been through a few filthy loft spaces, his hair completely unkempt, smelling of sweat and other unsavoury odours, and looking like he really doesn't want to stay to deliver the training because he is just too tired? One would hope not! Yet too often this is what the proud owner of a new, expensive CCTV system (or other security system) is presented with when the engineer presents the handover.

A well-presented handover begins like any other well-presented training session, with planning and preparation. The installer should know in advance when the system will be ready for handover, and should therefore plan ahead. It is important to know how many people will be involved in the use of the system and, where relevant, the level of authority of each of these people. The owner should be made aware of the need to have these people available to attend the handover training. Also let the owner know that the area must be 'controlled' during the handover; that is, interruptions should not occur, and it should be possible to demonstrate the operation of all equipment. For example, there is no point in trying to demonstrate VMD in a department store when it is full of customers. A mutually convenient day and time should be agreed, which may be done verbally, but it is much better if the arrangements are made in writing, therefore covering any 'misunderstandings'. Finally, when the engineer turns up for the handover session, he/she should appear well presented and looking prepared. Some installing companies insist that before beginning a handover the installer washes, puts on a shirt and tie and generally makes themselves presentable. In the case of larger installations it is more than likely that handover will not take place directly following the fixing of the last cable clip and sweeping of the control room floor area, so the engineer will be travelling to the site, either directly from home or the office, and should do so suitably dressed.

The amount of preparation for handover will of course be governed by the size of the system. The handover of a town-centre-sized system cannot be compared to that for a small system comprising two cameras, one monitor and a DVR in a local petrol station. Nevertheless, the same principles can still be applied to both situations.

Before beginning the handover, make sure that all documents are available and correct. These documents should include such things as user manuals, handover agreement, maintenance agreements, contact numbers and addresses, and copies of commissioning checklists.

Upon meeting the owner and any other persons involved in the handover, establish by means of questioning how much they know or do not know about CCTV. In the majority of cases people know very little about the operation of CCTV equipment; however, where a system has been upgraded, it might be that the installer is only required to explain the differences the upgrade may have made to the system operation and capability.

At the start of the handover, walk the owner around the system, showing them the location of every item of equipment, making them aware of the field(s) of view for each camera. Demonstrate the operation of the equipment, making sure that the customer regularly handles it for themselves during the handover. In fact, the owner should handle the equipment more than the engineer during this period.

Tape management, as outlined in Chapter 8, should be explained, as should the implications of the Data Protection Act 1998 (applicable in the UK). That is, that the owner of a CCTV system is legally responsible for the security of recorded material, and should any member of the public bring a formal complaint about recorded material being released into the public domain, either intentionally or otherwise, the CCTV system owner might well be brought to court. In the UK, the days of passing CCTV footage to TV companies for the purposes of entertainment are at an end, because any member of the public upon seeing themselves 'starring' in such a programme, or even appearing incidentally, could complain that it is a misuse of the system, and that it is in breach of their human rights.

In relation to maintenance, instruct the owner in the use of wash/wipe, de-misters, etc., as well as VCR head cleaning cassettes (Chapter 8). Inform them of the need to store tapes away from the magnetic fields which emanate from many items of electrical/electronic equipment. Also, where relevant, point out the health hazard associated with unshielded tape degaussers (bulk erasers); there is the possibility of these units upsetting the operation of heart pacemakers, and people who have one of these fitted should never operate this equipment.

Preventative maintenance

The BSIA Code of Practice for the Planning, Installation and Maintenance of Closed Circuit Television Systems; 109: Issue 2: October 1991 breaks preventative maintenance down into two levels: Level 1, which is basic- ally a full system test, and Level 2, which is a full system test plus an inspec- tion of all equipment associated with the installation.

A Level 1 visit must be made at least every twelve months, the first visit taking place within twelve months of the handover. The engineer should test the operation of every piece of equipment, including each wash/wipe facility, lights, PTZ, etc. The picture quality from each input should be examined, and careful attention paid to loss of quality due to such things as dirt on the housing windows or cobwebs. Also check the picture con- trast, colour quality and noise level, noting any possible deterioration and logging it for further investigation.

Compare the system operation against the original OR. Make sure that fields of view are unaltered, housing stop positions are correct, etc. Also make sure that the field of view for each camera has not become obstructed by objects which were not present when the system was installed – for instance, the large tree which was a mere sapling a few years previously, or the new lamp post which the council thought to site directly in front of the camera. Although it is not the job of the engineer to cut down the tree (or dismantle the lamp post!), it is his/her responsibility to inform the owner of the system of environmental changes which mean that the system is no longer able to perform to the OR.

A Level 2 visit, as defined by the BSIA, does not have a fixed frequency, but is to be agreed between the owner and the installing company. On the other hand, the sort of checks that this visit involves should be made at least every two years if not every year.

During a Level 2 visit, all of the system tests for a Level 1 visit are car- ried out. In addition, visual inspection of all equipment must be carried out. Camera housings, receiver units and P/T units must be opened and inspected for corroding of electrical terminals, water ingression and any other signs of wear or damage. P/T units should be greased, wiper blades replaced as necessary, heater performance verified, flexible cables checked for damage or wear, etc.

Camera towers or columns should be inspected for signs of damage or corrosion, as should wall-mounting brackets. In particular, the fixing points of towers and brackets should be inspected for signs of corrosion, fatigue and loose or missing bolts.

All lighting should be tested, and where the lights are not easily accessible it is suggested that the lamps are replaced as a matter of course whilst access equipment is in place. This is probably more cost-effective than having to return in the not too distant future to replace a defective lamp.

Any repair or replacement work that cannot be completed at the time of the maintenance visit should be recorded, and arrangements made for the work to be completed at the earliest opportunity. All maintenance work should be recorded and the document signed by both the engineer and the owner or a representative of the owner. A copy of the record should be left for the owner, and the original returned to the company office where it should be stored in a secure place for future reference.

Where maintenance or servicing involves the temporary disconnection of equipment which will leave a part of the system inoperative, a temporary disconnection notice should be completed and signed by both the engineer and the owner, or other responsible person.

With particular reference to VCRs, as discussed in Chapter 8, because of the wear and tear on the mechanism in a time-lapse machine, a full service should be carried out every twelve months. This will involve the removal of the machine from the site for a period of time, and it may be necessary to install a machine on temporary loan. In this case the engineer must compare the features on the loan machine with the one that it is replacing ensuring that either (a) the machine is capable of offering all of the features needed to maintain the system performance or (b) the customer is made aware of any VCR operations that are not available.

At all times the engineer must be careful not to degrade the performance of any part of the system by replacing defective components with parts of a lower specification or parts that are incompatible. For example, if a camera is found to be defective, the replacement should be a model that is of equal if not better specification. Similarly, if the hard disk in a digital recording system fails, replacing it with a lower-capacity unit will have a serious impact on the recording and archive period.

Corrective maintenance

Just as it is necessary to create accurate and legible preventative maintenance records, the same applies to any corrective work that is carried out on the system, either during routine visits, or in response to a breakdown call.

Copies of all corrective maintenance documents must be stored in a secure place in the system file at the company office. On the one hand this protects the company in the case of any redress from the owner who may in the future find reason to accuse the company of failing to carry out certain work. On the other hand the information in these records can prove helpful in the event of future problems with the system.

Fault location

This subject has to a large extent already been dealt with, having considered the fault diagnosis and repair techniques pertinent to the equipment under discussion in each chapter. However, we will now look at fault diagnosis from a *system* point of view.

A logical approach to fault diagnosis is essential if servicing is to be both sound and cost-effective. But what is logical fault diagnosis? Well, perhaps it is easier to begin by stating what it isn't: the engineer who, when faced with a problem, falls into a state of panic and throws all theory knowledge aside and instead resorts to replacing items of equipment at random until at last the problem goes away, is certainly not using a logical approach to fault diagnosis. Having personally observed both CCTV and intruder alarm trainees in this 'state', it has been interesting to note that in many cases not only is the theory thrown aside, but the test equipment remains safely stored in (unopened) cases. In some instances trainees have been seen to pull out and replace a complete cable run just to see if it is the cause of a problem, before having applied any form of test meter to the cable. This is certainly *not* a logical approach to fault diagnosis!

To locate a fault efficiently, an engineer would usually begin by giving careful consideration to the fault symptom(s). Television can be a very informative piece of equipment in this respect because each symptom displayed on the screen very often points to the fault area, as long as the engineer can recognize the symptoms. From the fault symptom, the engineer should be able to deduce which parts of the system may be responsible, and which parts can be ruled out.

The next step in logical fault diagnosis very often is to fall back onto experience. Here the engineer's mind is working somewhat like a computer database, searching though archive files for any previous experiences of the same symptoms on similar equipment. This does not mean to say that the cause of the trouble will definitely turn out to be the same, but where the engineer is aware of a common problem in a certain item, or combination of equipment, it makes sense to begin by testing in this area.

Where experience alone does not produce a rapid diagnosis, the engineer must change gear and begin to search his/her mental 'database' for the appropriate CCTV and/or electronics theory. The skill and ability of an engineer can usually be determined from how well they can do this because, as much as it is unsatisfactory to throw the theory aside, it can be equally unhelpful to pull down so much theory that the engineer becomes totally confused. Having the ability to apply the correct theory at just the right moment is a skill in itself.

It is one thing to have the knowledge, yet this may not be a lot of use unless one has access to the test equipment to enable voltage and waveform measurements to be taken. A calibrated, quality multimeter is the bare minimum that an engineer should carry. With this the engineer can verify the presence and value of a.c. and d.c. voltages and currents, as well as test for open circuit, short circuit and resistance faults. Remember that when measuring mains voltage potentials, only approved shrouded and fused test leads must be used.

For testing co-axial cables the multimeter resistance ranges are of limited use, and therefore a *time-domain reflectometer* is another very useful piece of test equipment for the service engineer, which can very quickly pay for itself by saving a lot of time and effort in trying to locate a fault along a length of co-axial cable. The TDR was discussed in Chapter 2 (Figure 2.32).

Because we are dealing with video signals it has to go without saying that the best way of testing the condition and level of these signals is to display them on an oscilloscope. And yet the CCTV industry is proving very reluctant to move in this direction for fault location. There are a number of reasons for this. First of all it is fair to say that lugging a scope around a site (possibly in the rain) is not the best thing for either the engineer or the scope. Secondly, a scope suitable for CCTV use will cost at least £400 brand new, and thus it is not viable to provide every engineer in a company of any size with such an item. But there is a third reason for not using oscilloscopes, and this is that many engineers are afraid of them and simply don't know how to operate one. This third reason is not a valid one for ignoring the value of an oscilloscope in locating certain difficult types of fault.

The truth is that the CCTV engineer does not have to be an expert at operating an oscilloscope, because he/she is not trying to display a wide range of signals. In practice the engineer usually wishes to display the same type of signal every time, this being one or two lines of a 1 Vpp CCIR television signal. Therefore, once the scope has been set to display this signal, it will never require anything more than possibly minor fine adjustment each time it is used for this purpose. Later in this chapter we shall look at some default settings for using an oscilloscope for CCTV applications.

Hand-held oscilloscopes are available and are much more convenient to use for CCTV system testing. However, the less expensive units tend to offer only a limited bandwidth and can have a slow screen refresh rate, making it difficult, if not impossible, to notice any rapid fluctuations in signal level or condition. This is not to say that all hand-held scopes suffer this problem, and devices such as that illustrated in Figure 13.3 are very effective and are ideally suited to CCTV applications. Being menu driven, scopes such as this very often include a 'Video' menu option which sets the vertical and horizontal scales such that the scope will display one or two lines of video information. The main drawback with units such as this is their relatively high cost compared with comparable bench models.

An alternative to using an oscilloscope to determine the condition of a video signal is to use any one of a number of hand-held video signal level indicators that are available. Compared to an oscilloscope these are far less expensive, cumbersome and delicate and are a worthwhile addition to any service engineer's test kit. An example of such an indicator is illustrated in Chapter 2; Figure 2.31. This version comes with its own signal generator, which brings us to our next point.

Figure 13.3 *A hand-held oscilloscope is far more suited to CCTV testing than a full-sized bench model. Units such as this incorporate rechargeable batteries, and have default settings to enable rapid display of analogue video waveforms*

Scoping or measuring video signals from cameras is not the best way of testing for some faults, and the engineer sometimes needs to know for sure that the signal being injected into a cable or item of equipment is a CCIR standard video signal. To this end, a small portable video signal generator is very useful for fault-finding in CCTV systems. The simplest of these puts out a black and white bar signal; more sophisticated versions produce a selection of coloured and black and white patterns, including perhaps the most useful which is the standard colour bar display, or staircase when viewed on an oscilloscope.

Oscilloscope default settings

One TV line has a time duration of 64 μs, and a peak-to-peak voltage of approximately 1 V (when terminated into 75 Ω). To display this waveform on an oscilloscope the volts/div control must be set to around 0.2 V, and the timebase speed adjusted for around 10 μs/div. If the signal level is correct, then a stable trace should be obtained with the trigger control set to the 'Auto' or 'Normal' position. These are the default settings, and once a scope has been adjusted to display a stable video signal waveform, if the controls are left in this position, then a stable trace should be displayed immediately a video signal is applied to the input.

The expected display for these settings when a grey scale video input is applied is illustrated in Figure 13.4.

Oscilloscopes differ between models, and some offer features that the CCTV engineer will never use; however, these features result in a lot more controls on the front, making them more difficult to set up. (They are actually no more difficult to set, but the additional controls give more opportunity for the unwary operator to become confused.) When choosing an oscilloscope for CCTV use, a basic dual-beam model with a minimum bandwidth of 20 MHz is an ideal choice. The bandwidth of an oscilloscope is important as, among other things, it determines the triggering ability. Anything below 20 MHz will struggle to display a stable video waveform, and models with a bandwidth of 30–40 MHz are generally a good compromise between quality and price (for CCTV applications).

For those who do not regularly use an oscilloscope, the following procedure should produce a stable, single beam display on the majority of models (refer to Figure 13.4).

1. Switch on
2. Adjust INTENSITY (Brightness) control to mid position
3. Set TRIGGER LEVEL control to AUTO (NORMAL) position
4. Adjust X and Y POSITION controls for centre position
5. Set channel 1 Y GAIN/AMPLITUDE control to the 0.2 V position (or nearest value on your scope)
6. Set Timebase control to 10 μs/div
7. Set TRIGGER SELECT switch to CH1 position
8. Set trigger switches to AC and + positions
9. Set INPUT COUPLING switches on channels 1 and 2 to AC position
10. Connect scope to the equipment under test.
11. Adjust FOCUS and INTENSITY controls for best definition.

Provided that the signal under test is good, then a stable display should be obtained. If the display is jittering or moving horizontally across the screen, try adjusting the trigger control, perhaps switching it to the MANUAL position. If in doubt as to whether or not the scope is set correctly, try applying a known good video signal rather than the one under test. A correct display will confirm both that the scope is set up correctly, and that the signal under test is indeed incorrect.

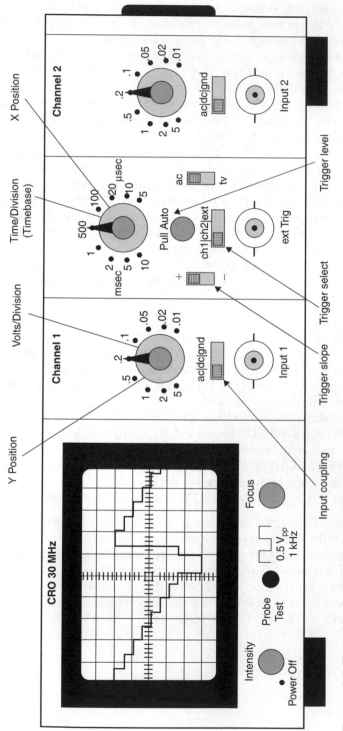

Figure 13.4 *Typical basic dual beam oscilloscope*

Glossary of CCTV terms

Some of the terms listed in this glossary may have more than one meaning, depending on the context in which each is taken. Where this is the case, the definitions given here relate to closed circuit television.

Alarm activation   A piece of equipment that is made to change its mode of operation by the activation of an alarm input. Typically a VCR, multiplexer or matrix switcher.

Algorithm   A mathematical term used to describe a procedure for resolving a problem.

Amplitude modulation (AM)   A transmission method where the signal to be sent is added to a much higher frequency carrier signal in such a way that the carrier amplitude is made to change in sympathy with the wanted signal. Used to transmit audio and video signals. Also used in magnetic video tape recording.

Analogue   A signal that is represented by changes in voltage level.

Aperture   The size of the hole produced by the iris. The size of the aperture determines the amount of light that will fall onto the pick-up device.

ARP   Address resolution protocol. A protocol which enables communication with a host using its MAC address in order to assign an IP address.

Artefact   In relation to digitized video signals, an artefact is any visible component in the picture display which is not a part of the original. This is rather like 'noise' in analogue displays; however, where large amounts of compression of the video signal has taken place, many artefacts may appear as a result of both losses and incorrect attempts by the decompression circuits to 'guess' at the picture content.

Aspect ratio   The ratio of the horizontal and vertical monitor screen dimensions. For CCTV this is 4:3.

Aspherical lens   A non-spherical lens that produces much less optical distortion than a conventional lens and, because of the way that the light travels through the glass, offers a higher light output for the same aperture setting.

Attenuation   Term used to describe a reduction in signal amplitude as it passes through a system or medium.

Automatic gain control (AGC)   Amplifier circuit that automatically changes its gain as the input signal level varies to maintain a constant output signal level.

Automatic iris (AI)   Iris that is controlled by the signal level in the camera. Changes in light input level result in changes in signal amplitude which are in turn used to adjust the iris to compensate.

Automatic level control (ALC)   A circuit that maintains a constant output signal level despite large changes in input signal level (see also AGC). In CCTV such a circuit may be used to control the lens iris.

Back focus   Mechanical adjustment of the position of the image device in a camera. Used to set the distance between the back of the lens and the pick-up to 12.5 mm for a CS mount lens, or 17.5 mm for a C mount lens.

Back light compensation (BLC)   Feature in a camera that causes the iris circuit to ignore the bright areas of the image and to open up sufficiently to allow the darker areas to be visible – albeit at the expense of the light areas.

Back porch   Period lasting 5.9 µs following each horizontal sync pulse. It is at black level to ensure that the electron beam in the CRT is cut off for the duration of the line flyback period.

Backbone   Main bus onto which all network devices connect. Each individual connection to the backbone is known as a segment. The backbone usually has a much broader bandwidth than the segments as it has to carry all of the traffic on the network.

Balanced cable   A two-wire method of signal transmission where the signal phases are such that unwanted interference signal are cancelled out at the receiving end.

Bandwidth   A term that is used to define the range of frequencies between the upper and lower cut-off points of a transmission system (measured at the −3 dB points). In CCTV there is a direct relationship between bandwidth and picture resolution.

Baud   Unit of measuring the data rate in a digital system, usually in bits per second (bps).

Bit   Derived from the term **bi**nary dig**it**. This is a single piece of data at either logic zero or logic 1.

Black level   In a video waveform, this refers to the voltage level that produces black level on the CRT or other display device. For a CCIR standard signal this is 0.3 V.

Blanking period   Times during both the horizontal and vertical scanning periods when the electron beam in the CRT is returning ('flying back') to the position for the next start of scan. During these periods the video signal waveform is held at the black level (0.3 V) to ensure that nothing is displayed on the screen. For a UK PAL transmission, the line or horizontal blanking period is 12 µs and the field or vertical blanking period is approximately 1 ms.

BNC   Co-axial cable connector commonly employed in CCTV installations. The term 'BNC' is derived from 'Bayonet–Neil–Concelman'.

Bridge A network device which connects two segments (similar to a switch, except that a switch can connect a number of pairs of segments). It is used to keep traffic on the two segments separate except when it is necessary to pass it over.

Bus Electrical conductor carrying one or more data signals. Many items of electronic equipment have internal bus lines in the form of copper tracks on the PCB, but in CCTV applications we are usually referring to external data bus lines such as RS 485, RS 422, RS 232, etc.

Byte Term used to define a group of 8 bits (binary digits) in a digital system. The term was derived because the earliest computers used only 8-bit words for communication, whereas modern processors commonly communicate using 32-bit words.

C mount Common screw thread developed for lenses used by the cine industry but adopted by the CCTV industry. A C mount lens has a distance of 17.5 mm between the image device and the back of the lens.

Cable equalizing Equipment added to a co-axial cable transmission system for the sole purpose of correcting for high-frequency losses in the cable.

Candela The unit of luminous intensity. One candela is the amount of light that is radiated in all directions from a black body that has been heated to a temperature equal to that at which platinum changes from a liquid to a solid state.

CCD (charge coupled device) A solid state device that is capable of storing small electrical charges. Originally intended for use as a digital store for computing applications, it has been adapted to store the output voltages from the photo diodes in a camera pick-up prior to them being passed on to the signal processing stages.

CCIR (Committée Consultatif Internationale des Radiocommunicationale) The English translation means 'Consultative Committee for International Radio'. This is the European body that has been responsible for the setting of Standards for television systems in Europe.

CCTV (closed circuit television) A television system that is not broadcast to air and therefore its images can only be accessed by persons with a connection to that system.

Characteristic impedance The impedance of a cable. This is a function of the inductive, capacitive and d.c. resistive properties of a cable and is usually quoted assuming a (theoretical) cable of infinite length. For CCTV applications, co-axial cable must have a characteristic impedance of $75\,\Omega$.

Charge An accumulation of electrons in a device or area. Electrical charge is measured in coulombs and is equal to an accumulation of 6.289×10^{18} electrons.

Chroma burst　A sample of the colour subcarrier signal. It is transmitted during the back porch period and comprises 10 cycles of the 4.43 MHz subcarrier for a PAL transmission, or up to 10 cycles of the 3.58 MHz subcarrier for NTSC transmissions. The burst signal serves as a form of sync signal for the colour decoding circuits in a television, monitor or VCR.

Chrominance　Term used to define the colour signal components in a television transmission.

CIF　Common interchange format. A set of standards which specify picture formats in pixel sizes.

Co-axial cable　An unbalanced cable comprising a core surrounded by a braided or solid screen. The two conductors are separated by an insulating material that is designed to act as a capacitive dielectric. The total inductive, capacitive and d.c. resistive properties of the cable produce an opposition to a.c. currents that is known as the characteristic impedance. For CCTV applications, co-axial cable must have a characteristic impedance of 75 Ω.

Colour temperature　The type of light produced by a source. It is a scientific measure of the wavelengths of light, stated in degrees Kelvin (K).

Composite sync　A signal that contains both the horizontal and vertical sync pulses but no video signal components.

Composite video　Signal containing the luminance, chrominance, sync and colour burst components.

Compression　With respect to digitized video signals, this refers to the removal of binary data containing information relating to the signal that is considered, for one reason or another, to be redundant. Compression is required to overcome the need for the large amounts of digital storage capacity associated with digitized video signals.

Cross-modulation　An effect whereby two high-frequency (r.f.) signals interact, resulting in an unwanted interference.

CS mount　The most common lens mounting type used in CCTV. It has a distance of 12.5 mm between the image device and the back of the lens. A CS mount lens cannot be used with a C mount camera.

CSMA/CD　Carrier sense multiple access/collision detection. Method of data transmission used in Ethernet whereby hosts access the data bus as and when required. When a data collision occurs, hosts re-send their data, and will continue to do so until the transmission is successful.

CVBS　See Composite video.

Dark current　Thermally induced currents in a camera pick-up device which produce an apparent signal output even in a total absence of incident light.

Data cable　Cable used for the transmission of digital signals. Typically twisted pair but may be co-axial.

Datagram   A data packet with a TCP header. Ethernet devices transmit data across a network by breaking it down into these smaller packets and re-assembling them at the receiving end.

dB (decibel)   Logarithmic unit used to define the ratio between two signal amplitudes. A change of 3 dB represents a doubling or halving of electrical power; a change of 6 dB represents a halving or doubling of electrical voltage.

Depth of field   The range (in distance) in front of the lens where objects remain in focus. This decreases when either the focal length or the aperture size is increased (F-number decreased).

DHCP   Dynamic host configuration protocol. A server application that is used to automatically assign IP addresses to hosts on a network.

Digital signal   A voltage signal used to represent the binary values of zero and one.

Direct drive (DD)   Term used to describe a CCTV camera lens having an electro-mechanical iris that is controlled via a changing d.c. voltage derived in the camera.

Distribution amplifier   A device containing a single input and a number of amplifiers which provide separate (isolated) outputs. Used for sending a single video signal to a number of pieces of equipment without affecting the 1 Vpp level.

DNS   Domain name service. A service which enables people to work with user-friendly names rather than IP addresses. The DNS server resolves (converts) the domain names into the IP addresses of the hosts to which users of the network wish to connect.

DOS (disk operating system)   Software used by the majority of computers to enable communications between the CPU and hardware devices such as hard disk and printer.

Duplex   When used with respect to CCTV multiplexers (MUX), this term refers to a unit that is able to offer simultaneous recording and replay (note that for analogue systems, two VCRs must be available).

DVR   Digital video recorder. Video recording equipment that employs a hard disk medium for the storage of video information.

EI (electronic iris)   A circuit in a CCTV camera that varies the amount of time that the pick-up device is active during any field period. The 'exposure' time is reduced as the light input increases, thus emulating the action of a mechanical iris. This feature enhances the performance of a camera that has a fixed iris lens fitted.

EIA (Electronics Industry Association)   The body that has been responsible for the setting of Standards for television systems in the USA, Canada and Japan; namely, the NTSC 525 line system.

Electromagnetic radiation   Electric field with an associated magnetic field travelling through free space at the speed of light in the form of waves. A wide range of frequencies, known as the electromagnetic spectrum, exist. Some of these are used for the purposes of radio signal transmission; others produce the visible light spectrum, whilst others are harmful to humans (i.e., X rays, Y rays[**G.1] and gamma rays).

Electron beam   Stream of electrons emitted from the cathode of a cathode ray tube (CRT).

EMI (electromagnetic interference)   Electromagnetic signals that enter a system or item of equipment and corrupt the wanted signals. EMI may be naturally occurring (i.e., lighting, solar flares, static electric charge), or man-made (i.e., radio transmissions, electric motors, electrical switch gear).

F-number   Figure used to denote the size of aperture in a lens. It is determined by the ratio of the focal length to the effective diameter of the lens/iris aperture. A small F-number indicates a large aperture and a high light throughput.

F-stop   Aperture setting on a camera lens. Each stop position represents a doubling or halving of light throughput.

Fast scan   A technique for transmitting CCTV signals digitally over the ISDN telephone network.

fc (foot candle)   Unit of light measurement, used primarily in North America. 1 foot candle (1 fc) is approximately equal to 10 lux.

Fibre-optic transmission   Signal transmission method that uses light pulses passing through a thin length of optically clear material. Capable of low loss transmission over long distances and is impervious to the effects of RFI and EMI.

Field   This contains one half of the video information in a television frame. The CCIR Standard has 50 fields per second, each field containing 312½ lines. The EIA Standard has 60 fields per second, each field containing 262.5 lines.

Field of view   The area that may be viewed through a lens. It is determined by the relationship between the angle of view and the distance between the object and the lens.

FIT (field interline transfer) chip   CCD image device that combines the technologies of the interline transfer and frame transfer chips, offering improved S/N ratio, low smear and good low-light sensitivity.

FM (frequency modulation)   A transmission method where the signal to be sent is added to a much higher frequency carrier signal in such a way that the carrier frequency is made to change in sympathy with the wanted signal amplitude. Used to transmit audio signals. Also used in magnetic video tape recording.

Focal length The distance, measured in millimetres, from the secondary principal point in the lens to the final focal point (at the image device). A short focal length produces a wide angle of view.

Format With respect to CCTV lenses and cameras, this describes the size of the active area of the pick-up device; typically ⅔″, ½″, ⅓″, or ¼″.

Frame One complete television picture made up from 625 (525 NTSC) line of information over two interlaced fields. The frame rate is 25 (30 NTSC) per second.

Frame store Commonly a digital memory device capable of storing the data relating to one complete digitized television frame.

Frequency Measurement of the number of times in one second that a cycle repeats itself. The unit of measurement is hertz (Hz) or, in the USA, cycles[**G.2].

Front porch Period lasting 1.4 μs prior to each horizontal sync pulse. It is at black level to ensure that the electron beam in the CRT is cut off for the duration of the line flyback period.

FTP File transfer protocol. A part of the TCP protocol suite that is used as a means of transferring files between two hosts.

Galvanometer A device that can convert electrical energy into mechanical energy without the mechanical complexity of a d.c. motor. Used to control the iris veins in a DD lens.

Gamma correction Modification to the shape of a luminance signal in a camera made in order to correct for non-linearity in a CRT phosphor response. In many CCTV cameras this is adjustable to compensate for differing monitors or lighting conditions.

Gateway A network device through which all data must be routed in order for it to be passed on to another network, either local or wide area. A gateway is normally a router.

Gen lock Means of maintaining sync between cameras in a CCTV system. A master sync signal is taken from either one camera or from a master generator and is fed to the 'Gen Lock' input on each camera.

Ground loop Circulating earth current formed by the screen on a co-axial cable. Such currents can cause rolling shadows on the picture (known as a hum bar), poor synchronization, horizontal tearing of the picture, or loss of telemetry functions.

Ground loop transformer A 1:1 transformer capable of passing video signal frequencies. Placed in series with a co-axial cable to break the direct earth connection between camera and monitor formed by the cable screen.

Hard disk Data storage device employed in all computers. Uses a solid metal disk onto which data is stored using magnetic recording.

Hardware address   Also known as the MAC (media access control) address or the Ethernet address. A unique address comprising a twelve-bit hexadecimal code, separated by hyphens, that is assigned to every network device. Used by applications such as DHCP to identify a host so that, for example, an IP address may be assigned.

Hertz (Hz)   Cycles per second. The unit by which frequency is measured.

Hop count   Normally set at 32, this determines the maximum number of network routers that the datagram may pass through before it is discarded as undeliverable. If there were no hop count, undelivered data would simply continue to pass around the Internet indefinitely, causing the Web to become congested with such data.

Host   An Ethernet network device that uses an IP address.

HTTP   Hypertext transfer protocol. A protocol that manages communications between the web browser and the web server and ensures that a requested folder, link, etc., is opened.

Hub   A simple network device used to construct a star network topology. A typical hub will have four or eight ports. Data on any one port is immediately connected to all of the other ports.

Hue   This term refers to the frequency of a light source: red, green blue, etc.

Hum bar   See Ground loop.

Illuminance   Measurement of light in lumens per square metre. The unit of measurement is the lux.

Illumination   Measurement of light coming from a secondary surface or source. The unit of measurement is the lux.

Imaging device   Device that is able to convert light energy into electrical energy. In modern CCTV cameras this will be a charge coupled device (CCD) chip.

Impedance (Z)   The opposition to current flow in an a.c. circuit, measured in ohms. It is the combined effect of the d.c. resistance and the inductive and/or capacitive reactances.

Infrared cut filter   A filter that blocks the passage of infrared light frequencies. Such filters are used in colour CCTV cameras to prevent the ingress of IR light that would otherwise result in incorrect colour signal production.

Infrared (IR) light   Frequencies of light that are just below those of visible light (wavelengths between 700 nm and 10 mm). All CCD image chips are sensitive to these frequencies and in many cases this can be used to an advantage.

Infrared pass filter   A filter that only allows infrared light frequencies to pass through. Such filters are placed in front of white light sources in the manufacture of IR lighting units. Typical wavelengths for CCTV lighting units are 715 nm and 850 nm.

Infrared spectrum   See Infrared (IR) light.

Intranet   A private network comprised entirely of private lines connecting the sites.

IP conflict   A situation that occurs on a network when two hosts are assigned the same IP address. When such a condition exists, neither host will be seen on the network until one of them is removed, or one of the IP addresses is changed.

IRE (Institute of Radio Engineers (of America))   A unit of measurement for a 1 Vpp video signal. It divides the signal between sync tip and peak white into 140 equal levels. For example, when 140 IRE = 1 Vpp video signal, the 0.3 V sync pulse level would be 42 IRE, etc.

Interlaced scanning   Method of producing a television picture by scanning each frame in two parts, first the odd television lines and then the even lines. Used to reduce picture flicker in CRT display units.

Interline transfer chip   Type of CCD image chip where the storage areas are adjacent to the photodiodes. This technique eliminates the vertical smearing problems associated with frame transfer chips, but the chip is less sensitive in terms of light input/signal output.

Internal sync   Horizontal and vertical synchronizing pulses are produced by the camera internally.

IP rating   The index of protection rating for an enclosure, e.g., IP 65. The first digit indicates the level of protection against ingression by solids; the second digit indicates the level of protection against ingression by liquids.

Iris   mechanism in a camera that governs the amount of light input.

ISDN (integrated services digital network)   A high-speed telephone transmission system for digital signals. 64 kB and 128 kB speeds are available.

Kell factor   Figure used when deriving the horizontal resolution for a television picture.

LAN (local area network)   Term used to describe a data communications network within a defined area, usually a single site.

LCD (liquid crystal display)   A device which uses the twisted nematic property of liquid crystal to control the light passing through in order to produce a visual display.

LED (light emitting diode)   Gallium arsenide diode that emits light when a current passes through its PN junction.

Lens   In CCTV, this term usually refers to a lens assembly which is an array of lenses with an iris mechanism. Its function is to gather light and focus it onto the pick-up device.

Light meter   Hand-held measuring device comprising of a photo pick-up and a meter. The meter is calibrated to display the light level in units of lux.

Line fed   The camera power is supplied via the co-axial video signal cable.

Line locked   Synchronization pulses produced by each camera are referenced to the a.c. mains frequency.

Luminance (Y)   Monochrome or black and white content of a video signal.

Lux   Unit of measurement of light.

MAC address   See Hardware address.

Matrix switcher   Equipment used to switch a number of cameras between one or more monitors.

Microwave transmission   For CCTV, this is a video signal transmission method using microwave radio signalling, usually in the 3 GHz to 10 GHz band. Greater range than infrared signal transmission.

Modal distortion   Signal distortion in fibre-optic cable caused by a multiplicity of light paths through the cable.

Modem   Term derived from its function, which is a **mo**dulator/**dem**odulator. Used to interface equipment having a digital output, and a conventional analogue telephone line.

Monochrome   A black and white television signal or system.

MPEG (Moving Picture Experts Group)   A body set up by the ISO in 1988 to devise standards for audio and video compression.

Multimode   Fibre-optic cable where the light is made to 'bounce' from side to side as it travels along. This cable type introduces modal distortion, but it is much less expensive than mono mode cable, where the light travels in a direct line along the cable length.

Multiplexing   Signal transmission method where two signals are transmitted on the same cable or carrier in such a way that they can be separated at the receiving end. The term also refers to equipment capable of processing video signals such that more than one image can be displayed on a screen, or recorded simultaneously (see MUX).

MUX   Abbreviation for a video multiplexer. A unit that processes video signals such that more than one image can be displayed on a screen, or recorded simultaneously.

NAT (network address translation)   A method that allows efficient use of IP addresses when transmitting data from one network to another across the Internet. A router or proxy server takes the datagrams on a private network and removes the local IP address, replacing it with another address before passing it over the Internet.

ND (neutral density) filter   A lens filter that affects all frequencies of light in the visible spectrum by the same amount, therefore causing an overall reduction in the light level entering the lens. It is used to simulate low-light conditions, forcing the iris to open and giving the best conditions for lens focus adjustment.

ND spot filter   A graduated filter at the centre of a lens. It has minimal effect when the iris is wide open; however, its effect increases as the iris closes. This type of filter prevents the aperture from becoming too small to control effectively under bright light conditions.

NetBEUI (netBIOS extended user interface)   See NetBIOS.

NetBIOS (network basic input output system)   The standard interface to networks for IBM PCs. Although relatively simple to set up and administer, it cannot be routed, which makes it inefficient in terms of data traffic, and is only viable for smaller networks.

NIC (network interface card)   Also known as the Ethernet card. Every host on an Ethernet network requires one of these to connect to the network (although devices such as CCTV cameras usually do not employ an actual card). The NIC performs the functions of organizing the data into frames, managing the transfer of these frames between hosts, and error correction.

Node   General term used to describe any device that is connected to a network (see also Host).

Noise   Electrical interference on the video signal. Usually manifest as grain (speckles) over the picture.

NTSC (National Television Standards Committee)   Committee that set the colour television system standards for North America and Japan.

NVR (network video recorder)   A digital video recorder that incorporates an Ethernet connection to permit both IP camera inputs and IP monitoring facility.

Octet   In relation to IP, an octet is an 8-bit word having 256 possible combinations. An IP version 4 (IPv4) address contains four octets, each separated by a dot.

Optical fibre   A transparent material along which light can be transmitted.

Oscilloscope   Displays electronic signals in a graphical form, enabling them to be measured and analysed.

Overscan   The monitor display is adjusted such that the electron beam scans farther than the edges of the screen, resulting in a loss of some picture information around the edges.

PAL (phase alternate line)   The most common system for the transmission of analogue colour television signals. The system maintains correct colour reproduction by cancelling out the effects of signal phase errors that occur during transmission.

Patchcord   A short length of flexible fibre-optic cable with a connector fitted to each end. Used to reconfigure a route between two pieces of equipment.

PCM (pulse code modulation)   Signal modulation method whereby the width of a continuous stream of square wave pulses is made to change in sympathy with the modulating signal.

Pelco D   A set of telemetry protocols devised by Pelco for CCTV applications. Pelco permits other manufacturers of CCTV equipment to use these protocols, which adds a much welcome uniformity to the industry.

Photo cell   A device, the electrical resistance of which is determined by the light falling upon it.

Photodiode   A diode in which the forward bias current is determined not only by the applied voltage but also by the light falling upon the PN junction.

Ping command   Computer command executed in DOS. Used to confirm that an IP addressable item of equipment is connected to, and communicating with, the IT network.

Pixel   Term derived from **picture elements**. A pixel is a single element of picture information – the greater the number of pixels, the greater the picture resolution.

Port   In relation to TCP/IP, a port is an address which defines the association between the data being transmitted and the applications on both the source and recipient PCs for which the data is intended. The source port identifies the application that sent the data, whilst the recipient port identifies the application for which it is intended. Each port has a specific number; for example, HTTP is assigned port 80.

Primary colours   The three colours (frequencies) of light – red, green and blue – that are perceived by the human eye and are integrated by the brain to derive all the colours of the visible light spectrum.

Progressive scan   A method of producing a picture display in some flat panel display devices whereby all TV scanning lines are produced simultaneously, rather than using traditional interlaced scanning.

PSTN (public switched telephone network)   In the UK, the original low bandwidth telephone network provided by the Post Office and later British Telecom. Intended only for speech transmission, it is now used for the transmission of digital signals; however, its low bandwidth makes it of little use for CCTV applications.

Pulse and bar generator   An item of test equipment that produces a continuous black, white and grey video test signal.

Quad Abbreviation for a unit that enables four camera signals to be displayed simultaneously on a monitor.

RAID (redundant arrays of independent disks) A method of creating a large disk capacity by grouping a number of hard disk drives into an array, with the array acting as a single drive. The method by which data is distributed across the disks provides a high degree of protection against data loss in the event of a disk drive failure.

Raster The blank white screen that results from the scanning action of the electron beam in a CRT before video information is applied.

Reflectance A figure which represents the ratio of light falling onto and returning from a surface, expressed as a percentage.

Reflected light Area illumination multiplied by reflectance.

Refraction Effect on a ray of light whereby as it passes through different medium its velocity alters, causing it to bend.

Reserved IP address An IP address that is set in a DHCP scope (by the IT administrator) to be assigned to the same host (and only that host) every time that host connects to the network.

Resolution In relation to the definition of a television picture, this is a measure of the smallest detail that may be discerned. The most common unit of measurement is TVL (television lines).

Router An intelligent network device that is on a par with a PC in terms of complexity and technology. It is used to intelligently forward data from one network to another by finding the quickest path, perhaps through a number of other routers.

Routing table A list of IP addresses of network hosts built up inside a router. When a host attempts to send data to a host on another network, the router will use its routing table to determine the shortest path by which to send the data.

RS port RS = recommended standard. Standard input/output connectors used for data communications. Common standards are RS 232, RS 422 and RS 485.

Scan coils Inductive coils placed around the neck of a CRT. Currents passing through these coils produce electromagnetic fields which interact with the electron beam, causing it to deflect both vertically and horizontally.

Scanning Rapid horizontal and vertical deflection of the electron beam in a CRT to produce a light output from the entire screen area (see raster).

SECAM (sequential couleur avec memoire – sequential colour with memory) Colour television broadcast system developed and used in France. It is not directly compatible with either PAL or NTSC systems.

Segment   A term used in networking to describe a data bus. A typical network will have a number of segments separated by routers or switches. Each segment may have a large number of hosts connected to it.

Server   A network administration device that provides services to the client PCs and other devices connected to the network. Such services include DHCP and DNS, mail services, web connection, firewall, application software, print services, file management, etc.

Silicon intensified target (SIT)   Type of camera pick-up device developed for use in very low light conditions.

Simplex   When used with respect to CCTV multiplexers (MUX), this term refers to a unit that can record or replay images, but not at the same time (unlike duplex units).

Slow scan   An early video signal transmission system that used the conventional PSTN telephone network. It employed analogue transmission and had a very slow refresh rate.

S/N ratio (signal-to-noise ratio)   Measurement of the amount of noise in a signal, expressed in decibels (dB). For video signals, any figure less than 40 dB will result in unacceptable amounts of noise (grain) in the picture.

Static IP address   An IP address that has been assigned to the host manually, rather than automatically via DHCP. One assigned, such IP addresses will remain unchanged, unless a person makes a manual adjustment.

STP cable (shielded (screened) twisted pair cable)   A twisted pair cable that has an outer screen to provide added protection against the effects of RFI/EMI.

Subnet mask   Data included in the datagram that is used to identify the network and host parts of the IP address. Subnetting makes it possible to independently route networks, so increasing the efficiency of the network by reducing network traffic, reducing the size of the routing tables, providing a simple way of isolating one network from another, and enhancing the ability to provide network security.

S-VHS (super video home system)   Analogue video recording format based on VHS, but offering much higher picture resolution (400 TVL as opposed to the 240 TVL for VHS). S-VHS recordings will not replay on conventional VHS machines.

Switch   An Ethernet device used to connect hosts for the duration of their communication. At first glance a switch appears similar to a hub; however, a switch creates a dedicated path between the hosts. This improves network performance because it means that, for much of the time, more than one pair of hosts can communicate at the same time.

Synchronizing pulses   Pulses added (at the camera) to the video signal. Used by the monitor to maintain correct scanning correlation between the camera and the monitor.

TCP/IP (transmission control protocol/Internet protocol)   Two protocols (sets of rules) used together for the transmission of data over Ethernet.

Telemetry   Signalling system used to control functions at a camera head such as pan, tilt, zoom and wash. May be analogue, but is now more commonly a digital system that employs decoders (site drivers) at each camera installation.

Telephoto lens   Correct term for a zoom lens. These lenses have a long focal length, giving them a high magnification but a narrow angle (field) of view.

Telnet   A virtual terminal service which permits a user to connect two network hosts and take full control over the remote host. All mouse and keyboard activity is passed directly to the remote host so that applications may be opened and run remotely.

Termination   With respect to CCTV, this term is frequently used to describe the 75 Ω termination impedance required at each end of a co-axial cable. Modern equipment employs automatic termination methods that detect the presence of a BNC connector at the socket; however, for many years a termination switch would be found close to the video output socket on monitors or other signal processing equipment.

Timebase corrector (TBC)   An item of equipment that is used to align the timing of the sync pulses from all cameras in a system, thus preventing picture roll during camera switching. In most cases the TBC is incorporated inside the MUX.

Time-lapse VCR   A video recorder (commonly S-VHS) that is capable of recording for very long periods on a standard three-hour cassette by operating a continuous record/pause action.

Topology   A term frequently used in networking to describe the wiring configuration for a particular network; e.g., Bus, Star.

Transducer   Any device that converts one form of energy into another. For example, a CCD chip converts light energy into electrical energy, whereas a CRT will do the opposite.

Triplex   When used with respect to CCTV multiplexers (MUX), this term refers to a unit that is able to offer simultaneous live monitoring, recording and replay (note that for analogue systems, two VCRs must be available).

Twisted nematic   A property of liquid crystal whereby the molecules form into helices, causing light to change polarity by 90°. When a voltage is applied across the crystal structure, the helices break down and light is able to pass through unaffected (see also LCD).

Twisted pair   Type of cable having two wires that are twisted together, producing a balanced effect whereby electrical interference signals are cancelled out. Capable of much greater transmission distances that co-axial cable types. In some cases the cable may have an outer screen.

UDP User datagram protocol. A connectionless protocol that simply transmits datagrams across the network to the recipient host where they may arrive out of order, and are not subject to any error checking or correction.

Underscan Switchable feature on a monitor that enables the entire picture to be seen.

Unshielded twisted pair (UTP) Twisted pair cable type having no outer screen (see Twisted pair).

URL (universal resource locator) Used with HTTP to identify the location of a network device. A typical URL looks something like http://www.sitename.com.

Vacuum tube Device such as a CRT, fluorescent display or camera pickup tube. Uses thermionic electron emission to produce a current flow between a heated cathode and an anode.

Varifocal Lens having a manually adjustable focal length, giving a degree of choice over the field of view.

Vertical streaking Bright vertical lines produced in some types of CCD chip under certain conditions.

VHS (video home system) Analogue video recording format developed for the domestic market and later adapted for CCTV applications in the form of time-lapse recorders. It is a low-resolution format – 240 TVL for colour and 300 TVL for monochrome recordings.

Video launch amplifier Amplifier used to correct for signal losses in exceptionally long lengths of cable. Can be tuned to give greater compensation at higher frequencies where the cable losses are most acute.

Video line corrector Equipment used to correct for uneven frequency losses in long cable runs. Similar in action to a video launch amplifier, but it may not have such a high gain.

Video motion detection (VMD) An alarm detection facility whereby the picture content from a camera is analysed to look for evidence of change (movement). Originally analogue, modern equipment employs digital analysis techniques.

Video signal See composite video.

Wavelength Measurement (in metres) of the propagation distance of one cycle of an electromagnetic wave. It is taken between any two adjacent points in a waveshape. Relates directly to the signal frequency.

Wavelet (compression) Method of compressing a digitized video signal in order to reduce the amount of data (file size) needed to restore each picture frame. Uses a mathematical algorithm known as wavelet transform.

White level Voltage level in a video signal that produces peak white. For a CCIR signal this is 1 V (0.7 V above the black level).

WINS (windows Internet naming service)   A service designed to work on simpler networks where there is no DHCP or DNS server. Such networks generally use NetBIOS, which is simple to set up and administer, but it is only suitable for small networks because it cannot be routed and, in terms of bandwidth, is very inefficient.

Y/C   Term used to describe the separate luminance (Y) and chrominance (C) signals. An S-VHS signal connector uses Y/C signal transmission where the Y and C are passed along separate screened cores within the cable. This results in improved picture quality because the two signals are unable to interfere with each other.

Zoom lens   See Telephoto lens.

Index